Mathematics for Industry

Volume 13

Aims & Scope

The meaning of “Mathematics for Industry” (sometimes abbreviated as MI or MfI) is different from that of “Mathematics in Industry” (or of “Industrial Mathematics”). The latter is restrictive: it tends to be identified with the actual mathematics that specifically arises in the daily management and operation of manufacturing. The former, however, denotes a new research field in mathematics that may serve as a foundation for creating future technologies. This concept was born from the integration and reorganization of pure and applied mathematics in the present day into a fluid and versatile form capable of stimulating awareness of the importance of mathematics in industry, as well as responding to the needs of industrial technologies. The history of this integration and reorganization indicates that this basic idea will someday find increasing utility. Mathematics can be a key technology in modern society.

The series aims to promote this trend by (1) providing comprehensive content on applications of mathematics, especially to industry technologies via various types of scientific research, (2) introducing basic, useful, necessary and crucial knowledge for several applications through concrete subjects, and (3) introducing new research results and developments for applications of mathematics in the real world. These points may provide the basis for opening a new mathematics-oriented technological world and even new research fields of mathematics.

More information about this series at http://www.springer.com/series/13254

Katsuki Fujisawa · Yuji Shinano · Hayato Waki
Editors

Optimization in the Real World

Toward Solving Real-World Optimization Problems

Springer

Editors
Katsuki Fujisawa
Kyushu University
Fukuoka
Japan

Hayato Waki
Kyushu University
Fukuoka
Japan

Yuji Shinano
Zuse Institute Berlin
Berlin
Germany

ISSN 2198-350X ISSN 2198-3518 (electronic)
Mathematics for Industry
ISBN 978-4-431-55419-6 ISBN 978-4-431-55420-2 (eBook)
DOI 10.1007/978-4-431-55420-2

Library of Congress Control Number: 2015946581

Springer Tokyo Heidelberg New York Dordrecht London

Printed on acid-free paper

Springer Japan KK is part of Springer Science+Business Media (www.springer.com)

Foreword

Optimization in the Real World is a challenging book title that catches one's interest but needs explanation. The reason is that the key words of the title have—depending on the background of the reader—various meanings and interpretations and the connotations are even more diverse.

To make it clear from the beginning: This is a mathematics book. It presents a collection of chapters that are based on lectures given at the "IMI Workshop on Optimization in the Real World" that took place during October 14–15, 2014, at the Institute of Mathematics for Industry (IMI), Kyushu University, Fukuoka, Japan.

The chapters in this volume are of two types. One is of a methodological nature. The chapters of this type belong to the interface between mathematics and computer science. More precisely, algorithmic advances in the areas of linear and mixed-integer programming are addressed as well as challenges that arise when extremely large problems are approached on advanced supercomputers. The second type deals with applications such as turbine allocation for offshore and onshore wind farms, battery control for smart grid nodes, or supply chain network design. These are examples of what "real world" means in this volume.

Optimization is, in everyday life, often considered an attempt to do or make things better than before or than others. The desire to make good use of scarce resources and to be fast and efficient seems to be built into the human genome. Mathematics takes this ambition to the limit by its approach to constructing mathematical models of issues that arise in technology, business, other sciences, or society, to define concepts of optimality and to design and implement algorithms for solving the problems that arise this way.

The goal is always to find a true and provable optimum, of course. But there may be obstacles. The problems may be too large or too complicated, the methods may not be sophisticated enough yet, the available computers may still be too slow or too small. That is where heuristics are employed with which very often provably good solutions can be found in reasonable time and where practical or theoretical experiments with new types of computer architecture are conducted to extend the reach of mathematical technology.

This is what the workshop at IMI was about and this is the range of issues to which the chapters in this volume contribute.

I had the privilege of participating in the wonderful environment that has been built up at Kyushu University with the aim of bringing modern mathematical tools to industry. IMI appears to be very successful and on a steady course forward. The workshop was one significant step to highlight what mathematics together with computer science can achieve today to support industry when important applications need good solutions.

Berlin, Germany
June 2015

Martin Grötschel

Preface

This book contains the post-proceedings of the international workshop "IMI Workshop on Optimization in the Real World—Toward Solving Real-World Optimization problems," which was held in Fukuoka, Japan, during October 14 and 15, 2014. Optimization is not only a scientific field in mathematics and computer science, it is also strongly connected with the real world, especially industrial activity. Many optimization problems in the real world are often not solvable because they are on a huge scale and/or contain other essential difficulties. However, some such optimization problems are becoming solvable through the recent development of computing and optimization technologies. The purpose of the workshop was to provide an opportunity to communicate with researchers who deal with optimization problems in the real world, and to stimulate novel and innovative development in optimization technology.

The chapters of this volume discuss the theory and applications of mixed-integer programming and scientific computation, and show the importance, usefulness, and powerfulness of current optimization technologies, in particular, mixed-integer programming and its remarkable applications. This collection is intended for students, academic researchers, and non-professionals working on optimization in industry.

This volume has been published through a peer-review process. We would like to thank all the chapter authors, Dr. Timo Berthold (ZIB), and Dr. Guillaume Sagnol (ZIB) for their cooperation in the editing of this volume.

Fukuoka, Berlin
June 2015

Katsuki Fujisawa
Yuji Shinano
Hayato Waki

Schedule of IMI Workshop on Optimization in the Real World—Toward Solving Real World Optimization Problems

Schedule of October 14
(Each Talk Consists 40 min. Including Question Time)

13:30–13:40	Yasuhide Fukumoto (IMI, Kyushu University)—Opening Remarks
13:40–14:20	Katsuki Fujisawa (IMI, Kyushu University)
14:30–15:10	Kengo Nakajima (University of Tokyo)
15:30–16:10	Tobias Achterberg (GUROBI Optimization)
16:20–17:00	Gerald Gamrath (ZIB)
17:10–17:50	Matteo Fischetti (University of Padova)
18:15	Banquet @ ZauoBBQ by Bus

Schedule of October 15
(Each Talk Consists 40 min. Including Question Time)

10:00–10:40	Emerson Escolar (Kyushu University)
10:50–11:30	Inken Gamrath (ZIB)
11:30–13:00	Lunch @ Tenten
13:00–13:40	Andrea Lodi (University of Bologna & IBM-Unibo Center of Excellence on Mathematical Optimization)
13:50–14:30	Takafumi Chida (Hitachi)
14:50–15:30	Ryohei Yokoyama (Osaka Prefecture University)
15:40–16:20	Tomoshi Otsuki (Toshiba)
16:45–17:45	Martin Grötschel[1] (ZIB)
18:30	Banquet of IMI Colloquium

[1]This talk was organized as a part of IMI Colloquium and held on Lecture Room L-1, 3F.

Contents

Advanced Computing and Optimization Infrastructure for Extremely Large-Scale Graphs on Post Peta-Scale Supercomputers

Katsuki Fujisawa, Toyotaro Suzumura, Hitoshi Sato, Koji Ueno, Yuichiro Yasui, Keita Iwabuchi and Toshio Endo

Abstract In this paper, we present our ongoing research project. The objective of this project is to develop advanced computing and optimization infrastructures for extremely large-scale graphs on post peta-scale supercomputers. We explain our challenge to Graph 500 and Green Graph 500 benchmarks that are designed to measure the performance of a computer system for applications that require irregular memory and network access patterns. The 1st Graph500 list was released in November 2010. The Graph500 benchmark measures the performance of any supercomputer performing a BFS (Breadth-First Search) in terms of traversed edges per second (TEPS). We have implemented world's first GPU-based BFS on the TSUBAME 2.0 supercomputer at Tokyo Institute of Technology in 2012. The Green Graph 500 list collects TEPS-per-watt metrics. In 2014, our project team was a winner of the 8th Graph500 benchmark and 3rd Green Graph 500 benchmark. We also present our par-

K. Fujisawa (✉) · Y. Yasui
Institute of Mathematics for Industry, Kyushu University, 744 Motooka, Nishi-ku Fukuoka 819-0395, Japan
e-mail: fujisawa@imi.kyushu-u.ac.jp

Y. Yasui
e-mail: y-yasui@imi.kyushu-u.ac.jp

T. Suzumura
University College Dublin, Belfield, Dublin 4, Ireland
e-mail: toyotaro.suzumura@ucd.ie

H. Sato · T. Endo
Global Scientific Information and Computing Center, Tokyo Institute of Technology, 2-12-1 O-okayama, Meguroku, Tokyo 152-8550, Japan
e-mail: hitoshi.sato@gsic.titech.ac.jp

T. Endo
e-mail: endo@is.titech.ac.jp

K. Ueno · K. Iwabuchi
Department of Mathematical and Computing Sciences, Tokyo Institute of Technology, 2-12-1 O-okayama, Meguroku, Tokyo 152-8550, Japan
e-mail: ueno.k.ac@m.titech.ac.jp

K. Iwabuchi
e-mail: iwabuchi.k.ab@m.titech.ac.jp

K. Fujisawa et al. (eds.), *Optimization in the Real World*, Mathematics for Industry 13, DOI 10.1007/978-4-431-55420-2_1

allel implementation for large-scale SDP (SemiDefinite Programming) problem. We solved the largest SDP problem (which has over 2.33 million constraints), thereby creating a new world record. Our implementation also achieved 1.713 PFlops in double precision for large-scale Cholesky factorization using 2,720 CPUs and 4,080 GPUs on the TSUBAME 2.5 supercomputer.

Keywords Graph analysis · Breadth-first search · Optimization problem · High performance computing · Supercomputer · Big data

1 Introduction

The objective of many ongoing research projects in high performance computing (HPC) areas is to develop an advanced computing and optimization infrastructure for extremely large-scale graphs on the peta-scale supercomputers. The extremely large-scale graphs that have recently emerged in various application fields, such as transportation, social networks, cyber-security, and bioinformatics, require fast and scalable analysis (Fig. 1). The number of vertices in the graph networks has grown from billions to trillions and that of the edges from hundreds of billions to tens of trillions (Fig. 2). For example, a graph that represents the interconnections of all the neurons of the human brain has over 89 billion vertices and over 100 trillion edges. To analyze these extremely large-scale graphs, we require a new generation exascale supercomputer, which will not appear until the 2020s, and therefore, we propose a new framework of software stacks for extremely large-scale graph analysis

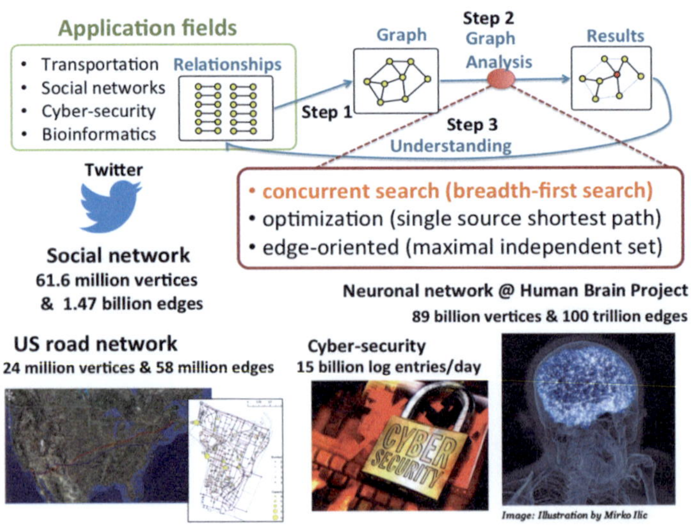

Fig. 1 Graph analysis and its application fields

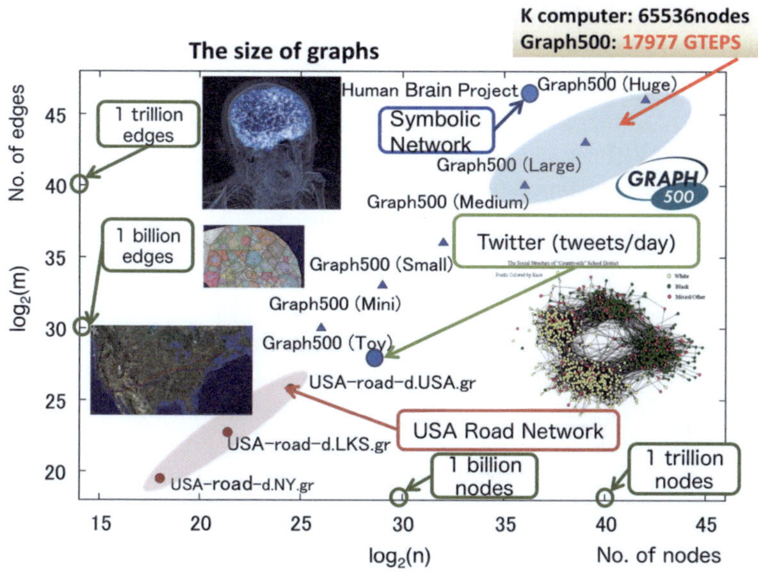

Fig. 2 Size of graphs in various application fields and Graph500 benchmark

systems, such as parallel graph analysis and optimization libraries on multiple CPUs and GPUs, hierarchal graph stores using non-volatile memory (NVM) devices, and graph processing and visualization systems.

We have a research team that joins the JST (Japan Science and Technology Agency) CREST (Core Research for Evolutional Science and Technology) post-Peta High Performance Computing project.[1] The objective of our researches for the JST CREST project is to develop advanced computing and optimization infrastructures for extremely large-scale graphs on post peta-scale supercomputers. In this paper, we explain our ongoing research project and show its remarkable results.

2 Graph500 and Green Graph500 Benchmarks

The Graph500[2] and Green Graph 500[3] benchmarks are designed to measure the performance of a computer system for applications that require irregular memory and network access patterns. Following its announcement in June 2010, the Graph500 list was released in November 2010, since when it has been updated semiannually. The Graph500 benchmark measures the performance of any supercomputer performing

[1] http://www.graphcrest.jp/eng/.

[2] http://www.graph500.org.

[3] http://green.graph500.org.

a breadth-first search (BFS) in terms of traversed edges per second (TEPS). The detailed instructions of the Graph500 benchmark are described as follows:

1. **Step1: Edge List Generation**
   First, the benchmark generates an edge list of an undirected graph with $n(=2^{SCALE})$ vertices and $m(=n \cdot edge\_factor)$ edges;
2. **Step2: Graph Construction**
   The benchmark constructs a suitable data structure, such as CSR (Compressed Sparse Row) graph format, for performing BFS from the generated edge list;
3. **Step3: BFS**
   The benchmark performs BFS to the constructed data structure to create a BFS tree. Graph500 employs TEPS (Traversed Edges Per Second) as a performance metric. Thus, the elapsed time of a BFS execution and the total number of processed edges determine the performance of the benchmark;
4. **Step4: Validation**
   Finally, the benchmark verifies the results of the BFS tree. Note that the benchmark iterates Step3 and Step4 64 times from randomly selected start points, and the median value of the results is adopted as the score of the benchmark.

We implemented the world's first GPU-based BFS on the TSUBAME 2.0 supercomputer at the Tokyo Institute of Technology and gained fourth place in the fourth Graph500 list in 2012. The rapidly increasing number of these large-scale graphs and their applications has attracted significant attention in recent Graph500 lists (Fig. 2).

In 2013, our project team gained first place in both the big and small data categories in the second Green Graph 500 benchmarks. The Green Graph 500 list collects TEPS-per-watt metrics. Our other implementation, which uses both DRAM and NVM devices and whose objective is to analyze extremely large-scale graphs that exceed the DRAM capacity of the nodes, which gained fourth place in the big data category in the second Green Graph500 list. In 2014, our project team was a winner of the 8th Graph500 (Fig. 4) and the 3rd Green Graph500 benchmarks (Fig. 5). Figure 3 shows our major achievements in Graph500 benchmark, which are mentioned in this Section.

As we have mentioned in this Section, our project team have challenged the Graph500 and Green Graph500 benchmarks, which are designed to measure the performance of a computer system for applications that require irregular memory and network access [6–8, 12, 13, 16–18]. We briefly explain four major papers of our research projects for Graph500 and Green Graph500 benchmarks.

1. "Highly Scalable Graph Search for the Graph500 Benchmark" [13]
   We found that the provided reference implementations are not scalable in a large distributed environment. We devised an optimized method based on 2D partitioning and other methods such as communication compression and vertex sorting. Our optimized implementation can handle BFS of a large graph with 2^{36} (68.7 billion vertices) and 2^{40} (1.1 trillion) edges in 10.58 seconds while using 1366 nodes and 16,392 CPU cores on the TSUBAME 2.0 supercomputer at Tokyo Institute of Technology. This performance corresponds to 103.9 GE/s. We also

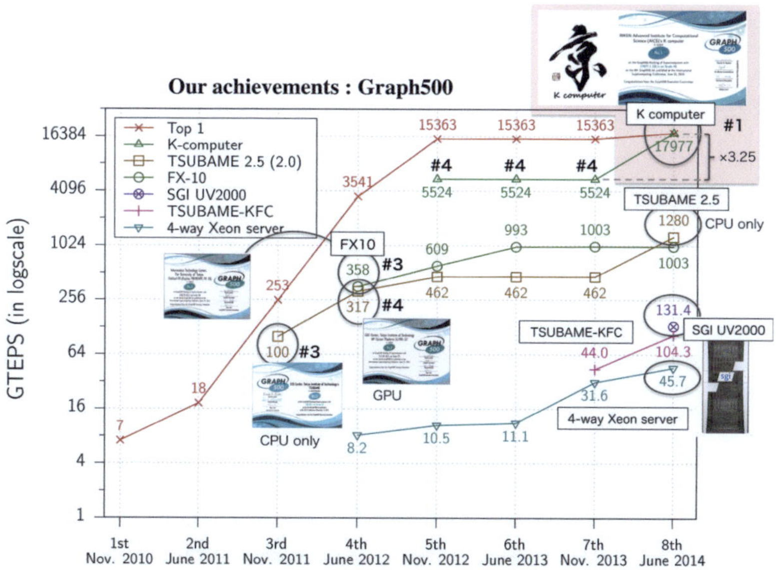

Fig. 3 Our major achievements in Graph500 benchmark

Fig. 4 Our project team were awarded the first place in the 8th Graph500 benchmark

studied the performance characteristics of our optimized implementation and reference implementations on a large distributed memory supercomputer with a Fat-Tree-based Infiniband network.

2. "NUMA-optimized Parallel Breadth-first Search on Multicore Single-node System" [16]

   Previous studies [1, 2] have proposed hybrid approaches that combine a well-known *top-down* algorithm and an efficient *bottom-up* algorithm for large fron-

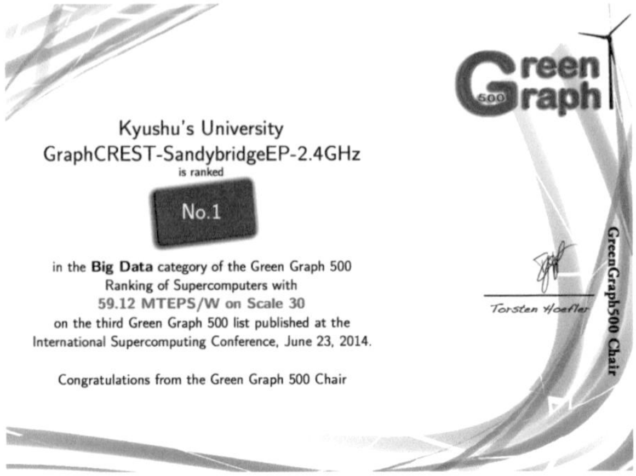

Fig. 5 Our project team were awarded the first place in the 3rd Green Graph500 benchmark

tiers. This reduces some unnecessary searching of outgoing edges in the BFS traversal of a small-world graph, such as a Kronecker graph. In this paper, we describe a highly efficient BFS using column-wise partitioning of the adjacency list while carefully considering the non-uniform memory access (NUMA) architecture. We explicitly manage the way in which each working thread accesses a partial adjacency list in local memory during BFS traversal. Our implementation has achieved a processing rate of 11.15 billion edges per second on a 4-way Intel Xeon E5-4640 system for a scale-26 problem of a Kronecker graph with 2^{26} vertices and 2^{30} edges. Not all of the speedup techniques in this paper are limited to the NUMA architecture system. With our winning Green Graph500 submission of June 2013, we achieved 64.12 GTEPS per kilowatt hour on an ASUS Pad TF700T with an NVIDIA Tegra 3 mobile processor.

3. "Fast and Energy-efficient Breadth-first Search on a single NUMA system" [18]
Our previous nonuniform memory access (NUMA)-optimized BFS [16] reduced memory accesses to remote RAM on a NUMA architecture system; its performance was 11 GTEPS (giga TEPS) on a 4-way Intel Xeon E5-4640 system. Herein, we investigated the computational complexity of the bottom-up, a major bottleneck in NUMA-optimized BFS. We clarify the relationship between vertex out-degree and bottom-up performance. In November 2013, our new implementation achieved a Graph500 benchmark performance of 37.66 GTEPS (fastest for a single node) on an SGI Altix UV1000 (one-rack) and 31.65 GTEPS (fastest for a single server) on a 4-way Intel Xeon E5-4650 system. Furthermore, we achieved the highest Green Graph500 performance of 153.17 MTEPS/W (mega TEPS per watt) on an Xperia-A SO-04E with a Qualcomm Snapdragon S4 Pro APQ8064.
4. "NVM-based Hybrid BFS with Memory Efficient Data Structure" [7]
We introduce a memory efficient implementation for the NVM-based Hybrid BFS algorithm that merges redundant data structures to a single graph data structure, while offloading infrequent accessed graph data on NVMs based on the detailed analysis of access patterns, and demonstrate extremely fast BFS execution for

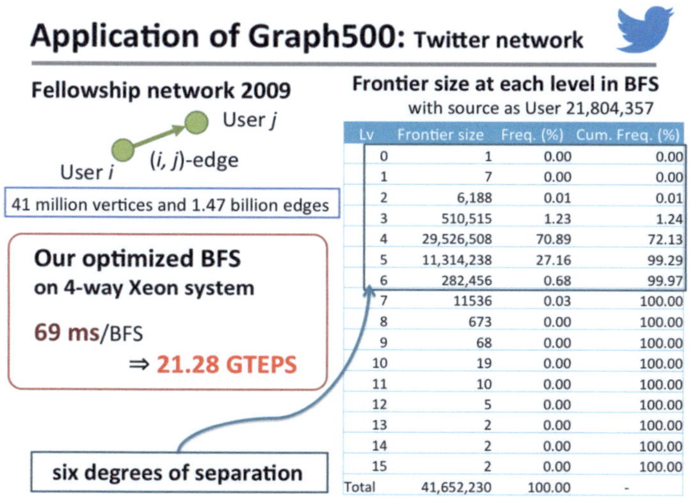

Fig. 6 Application of Graph500 benchmarks

large-scale unstructured graphs whose size exceed the capacity of DRAM on the machine. Experimental results of Kronecker graphs compliant to the Graph500 benchmark on a 2-way INTEL Xeon E5-2690 machine with 256 GB of DRAM show that our proposed implementation can achieve 4.14 GTEPS for a SCALE31 graph problem with 2^{31} vertices and 2^{35} edges, whose size is 4 times larger than the size of graphs that the machine can accommodate only using DRAM with only 14.99 % performance degradation. We also show that the power efficiency of our proposed implementation achieves 11.8 MTEPS/W. Based on the implementation, we have achieved the 3rd and 4th position of the Green Graph500 list (2014 June) in the Big Data category.

Figure 6 shows an application of the Graph500 benchmark. We slightly modified the source code for the Graph500 benchmark, which was applied to making a BFS tree of the Twitter Fellowship Network 2009. It takes only about 70 ms to make a BFS tree from a root node, although this graph has 41 million vertices and 1.47 billion edges.

3 High-Performance Computing for Mathematical Optimization Problems

We also present our parallel implementation for large-scale mathematical optimization problems [3–5, 14, 15]. In the last decade, mathematical optimization programming (MOP) problems have been intensively studied in both their theoretical and practical aspect in a wide range of fields, such as combinatorial optimization, structural optimization, control theory, economics, quantum chemistry, sensor network location, data mining, and machine learning. The semidefinite programming (SDP) problem is a predominant problem in mathematical optimization. The primal-dual interior-point method (PDIPM) is one of the most powerful algorithms for solving SDP problems, and many research groups have employed it for developing software packages. However, two well-known major bottleneck parts (the generation of the Schur complement matrix (SCM) and its Cholesky factorization) exist in the algorithmic framework of PDIPM. These two parts where bottlenecks occur are called ELEMENTS and CHOLESKY, respectively. The standard-form SDP has the following primal-dual form.

$$
\begin{aligned}
\mathscr{P} &: \text{minimize } \textstyle\sum_{k=1}^{m} c_k x_k \\
&\quad \text{subject to } X = \textstyle\sum_{k=1}^{m} F_k x_k - F_0, \quad X \succeq O. \\
\mathscr{D} &: \text{maximize } F_0 \bullet Y \\
&\quad \text{subject to } F_k \bullet Y = c_k \ (k = 1, \ldots, m), \quad Y \succeq O.
\end{aligned} \tag{1}
$$

We denote by $\mathbb{S}^n$ the space of $n \times n$ symmetric matrices. The notation $X \succeq O$ ($X \succ O$) indicates that $X \in \mathbb{S}^n$ is a positive semidefinite (positive definite) matrix. The inner-product between $U \in \mathbb{S}^n$ and $V \in \mathbb{S}^n$ is defined by $U \bullet V = \sum_{i=1}^{n} \sum_{j=1}^{n} U_{ij} V_{ij}$.

In most SDP applications, it is common for the input data matrices $F_0, \ldots, F_m$ to share the same diagonal block structure $(n_1, \ldots, n_h)$. Each input data matrix F_k $(k = 1, \ldots, m)$ consists of sub-matrices in the diagonal positions as follows:

$$
F_k = \begin{pmatrix} F_k^1 & O & O & O \\ O & F_k^2 & O & O \\ O & O & \ddots & O \\ O & O & O & F_k^h \end{pmatrix}
$$

$$
\text{where } F_k^1 \in \mathbb{S}^{n_1}, F_k^2 \in \mathbb{S}^{n_2}, \ldots, F_k^h \in \mathbb{S}^{n_h}.
$$

Note that $\sum_{\ell=1}^{h} n_\ell = n$ and the variable matrices X and Y share the same block structure. We define $n_{\max}$ as $\max\{n_1, \ldots, n_h\}$. For the blocks where $n_\ell = 1$, the constraints of positive semidefiniteness are equivalent to the constraints of the nonnegative orthant. Such blocks are sometimes called linear programming (LP) blocks.

The size of a given SDP problem can be approximately measured in terms of four metrics.

1. m: the number of equality constraints in the dual form $\mathscr{D}$ (which equals the size of the SCM)
2. n: the size of the variable matrices X and Y
3. $n_{\max}$: the size of the largest block of input data matrices
4. nnz: the total number of nonzero elements in all data matrices

We denote the time complexities of ELEMENTS and CHOLESKY by $O(mn^3 + m^2n^2)$ and $O(m^3)$, respectively.

We have developed a new version of the semidefinite programming algorithm parallel version (SDPARA), which is a parallel implementation on multiple CPUs and GPUs for solving extremely large-scale SDP problems that have over a million constraints [3, 5]. SDPARA can automatically extract the unique characteristics from an SDP problem and identify the bottleneck. When the generation of SCM becomes a bottleneck part, SDPARA can attain high scalability using a large quantity of CPU cores and some techniques for processor affinity and memory interleaving. SDPARA can also perform parallel Cholesky factorization using thousands of GPUs and techniques to overlap computation and communication if an SDP problem has over two million constraints and Cholesky factorization constitutes a bottleneck. We demonstrated that SDPARA is a high-performance general solver for SDPs in various application fields through numerical experiments at the TSUBAME 2.5 supercomputer, and we solved the largest SDP problem (which has over 2.33 million constraints), thereby creating a new world record. Our implementation also achieved 1.713 PFlops in double precision for large-scale Cholesky factorization using 2,720 CPUs and 4,080 GPUs (Fig. 7) [5].

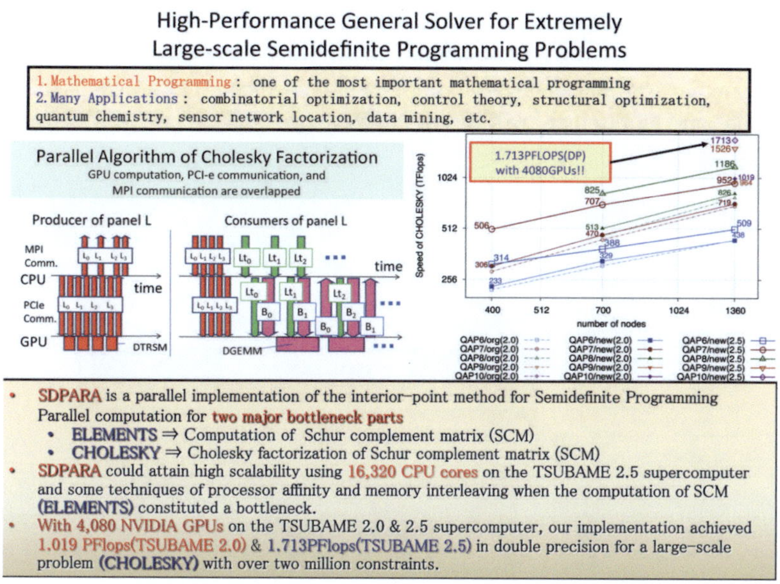

Fig. 7 SDPARA and its performance on TSUBAME 2.0 & 2.5 supercomputer

4 Software Stacks for Extremely Large-Scale Graph Analysis System and Future Plans

In this paper, we finally propose new software stacks for an extremely large-scale graph analysis system (Fig. 8), which are based on our current ongoing research studies above.

1. Hierarchal Graph Store: We propose a hierarchal graph stores and process extremely large-scale graphs with minimum performance degradation by carefully considering the data structures of a given graph and the access patterns to both DRAM and NVM devices. We have developed an extended memory software stack for supporting extreme-scale graph computing. Utilizing emerging NVM devices as extended semi-external memory volumes for processing extremely large-scale graphs that exceed the DRAM capacity of the compute nodes, we design highly efficient and scalable data offloading techniques, PGAS-based I/O abstraction schemes, and optimized I/O interfaces to NVMs.
2. Graph Analysis and Optimization Library: Large-scale graph data are divided between multiple nodes, and then, we perform graph analysis and search algorithms, such as the BFS kernel for Graph500, on multiple CPUs and GPUs. Implementations, including communication-avoiding algorithms and techniques for overlapping computation and communication, are needed for these libraries. Finally, we can make a BFS tree from an arbitrary node and find a shortest path between two arbitrary nodes on extremely large-scale graphs with tens of trillions of nodes and hundreds of trillions of edges.

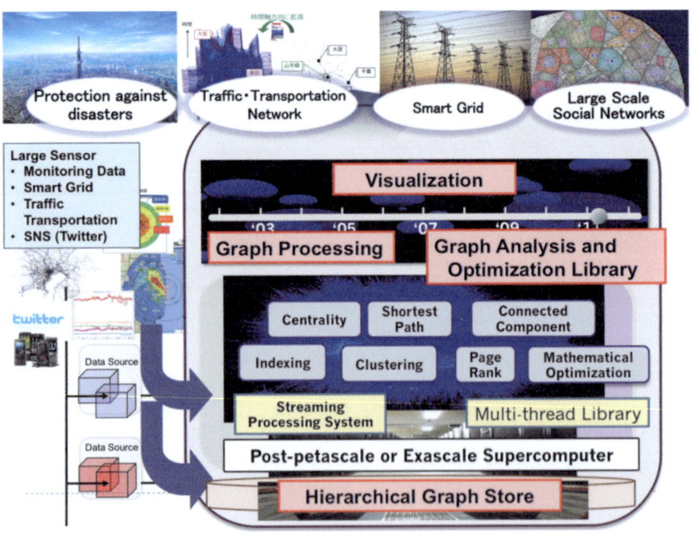

Fig. 8 Software stacks for extremely large-scale graph analysis system

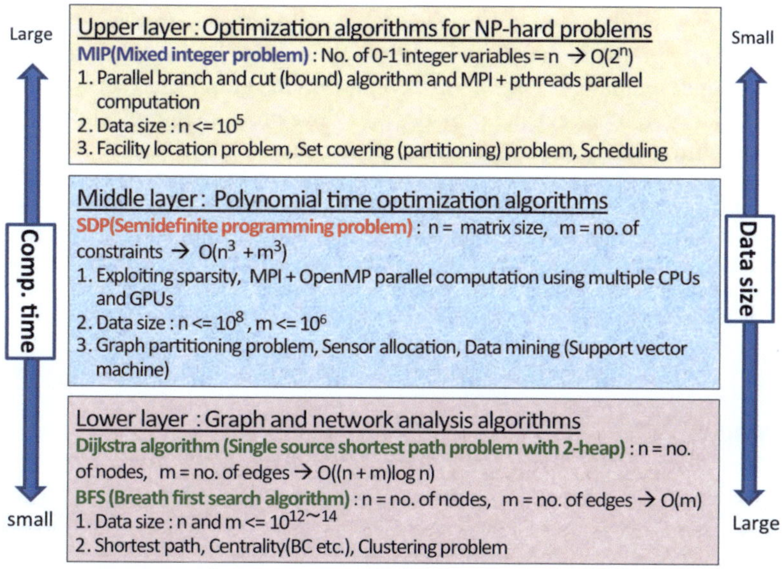

Fig. 9 Three layers of graph analysis and optimization library

3. Graph Processing and Visualization: We aim to perform an interactive operation for large-scale graphs with hundreds of million of nodes and tens of billion of edges.

We focus on the graph analysis and optimization library, which are illustrated in Fig. 8. Figure 9 shows three algorithmic layers of graph analysis and optimization library. We classify many optimization algorithms into three layers according to both of the computation time that we need to solve and the data size of the optimization problem. We have developed parallel software packages for many optimization problems categorized into these three algorithmic layers. The upper layer contains optimization algorithms for NP-hard problems. The most typical and important optimization algorithm in this layer is a branch and cut (bound) algorithm for the mixed integer problem (MIP). We have collaborated with ZIB (Zuse Institute Berlin) in developing and evaluating parallel (MPI + pthread) software package for solving MIPs [9–11]. The middle and lower layer contains interior-point algorithms for SDP problems and BFS for graph analysis, respectively. We'll release all the software packages developed in out projects until March, 2017 from the Web site.[4] We have started the research project for developing the Urban OS (Operating System) and implementing it on a large city (Fukuoka, Japan) from 2013.[5] The Urban OS gathers big data sets of people and transportation movements by utilizing different sensor technologies and storing them to the cloud storage system. As mentioned in this

[4] http://www.graphcrest.jp/eng/.

[5] http://coi.kyushu-u.ac.jp/en/.

paper, we have another research project whose objective is to develop advanced computing and optimization infrastructures for extremely large-scale graphs on post peta-scale supercomputers. The Urban OS employs the graph analysis system developed by this research project and provides a feedback to a predicting and controlling center to optimize many social systems and services.

Acknowledgments This research project was supported by the Japan Science and Technology Agency (JST), the Core Research of Evolutionary Science and Technology (CREST), the Center of Innovation Science and Technology based Radical Innovation and Entrepreneurship Program (COI Program), and the TSUBAME 2.0 & 2.5 Supercomputer Grand Challenge Program at the Tokyo Institute of Technology.

References

1. Beamer, S., Asanović, K., Patterson, D.A.: Searching for a parent instead of fighting over children: a fast breadth-first search implementation for Graph500. Berkeley, CA: EECS Department, University of California, UCB/EECS-2011-117 (2011)
2. Beamer, S., Asanović, K., Patterson, D.A.: Direction-optimizing breadth-first search. In: Proceedings of the ACM/IEEE International Conference on High Performance Computing, Networking, Storage and Analysis (SC12), IEEE Computer Society (2012)
3. Fujisawa, K., Endo, T., Sato, H., Yamashita, M., Matsuoka, S., Nakata, M.: High-performance general solver for extremely large-scale semidefinite programming problems. In: Proceedings of the 2012 ACM/IEEE Conference on Supercomputing, SC'12 (2012)
4. Fujisawa, K., Endo, T., Sato, H., Yasui, Y., Matsuzawa, N., Waki, H.: Peta-scale general solver for semidefinite programming problems with over two million constraints, SC13 regular, electronic, and educational poster. In: International Conference for High Performance Computing, Networking, Storage and Analysis 2013 (SC2013) (2013)
5. Fujisawa, K., Endo, T., Yasui, Y., Sato, H., Matsuzawa, N., Matsuoka, S., Waki, H.: Peta-scale general solver for semidefinite programming problems with over two million constraints. In: The 28th IEEE International Parallel & Distributed Processing Symposium (IPDPS 2014) (2014)
6. Iwabuchi, K., Sato, H., Mizote, R., Yasui, Y., Fujisawa, K., Matsuoka, S.: Hybrid BFS approach using semi-external memory. In: International Workshop on High Performance Data Intensive Computing (HPDIC2014) in Conjunction with IEEE IPDPS 2014 (2014)
7. Iwabuchi, K., Sato, H., Yasui, Y., Fujisawa, K., Matsuoka, S.: NVM-based hybrid BFS with memory efficient data structure. In: The proceedings of the IEEE BigData2014 (2014)
8. Iwabuchi, K., Sato, H., Yasui, Y., Fujisawa, K.: Performance analysis of hybrid BFS approach using semi-external memory, SC 2013 regular, electronic, and educational poster. In: International Conference for High Performance Computing, Networking, Storage and Analysis (SC2013) (2013)
9. Koch, T., Martin, A., Pfetsch, M.E.: Progress in Academic Computational Integer Programming, Facets of Combinatorial Optimization—Festschrift for Martin Grötschel, pp. 483–506. Springer (2013)
10. Koch, T., Ralphs, T., Shinano, Y.: Could we use a millon cores to solve an integer program? Math. Methods Oper. Res. **76**, 67–93 (2012)
11. Shinano, Y., Achterberg, T., Berthold, T., Heinz, S., Koch, T.: ParaSCIP: a parallel extension of SCIP. In: Competence in High Performance Computing 2010, pp. 135–148. Springer (2012)
12. Suzumura, T., Ueno, K., Sato, H., Fujisawa, K., Matsuoka, S.: A performance characteristics of Graph500 on large-scale distributed environment. In: The proceedings of the 2011 IEEE International Symposium on Workload Characterization (2011)

13. Ueno, K., Suzumura, T.: Highly scalable graph search for the Graph500 benchmark. In: HPDC 2012 (The 21st International ACM Symposium on High-Performance Parallel and Distributed Computing), Delft, Netherlands (2012)
14. Yamashita, M., Fujisawa, K., Fukuda, M., Kobayashi, K., Nakata, K., Nakata, M.: Latest developments in the SDPA family for solving large-scale SDPs. In: Anjos, M.F., Lasserre, J.B. (eds.), Handbook on Semidefinite, Conic and Polynomial Optimization, International Series in Operations Research & Management Science, Chapter 24 (2011)
15. Yamashita, M., Fujisawa, K., Fukuda, M., Nakata, K., Nakata, M.: Parallel solver for semidefinite programming problem having sparse Schur complement matrix. In: ACM Transactions on Mathematical Software, vol. 39, Number 12 (2012)
16. Yasui, Y., Fujisawa, K., Goto, K.: NUMA-optimized parallel breadth-first search on multicore single-node system. In: The Proceedings of the IEEE BigData2013 (2013)
17. Yasui, Y., Fujisawa, K., Goto, K., Kamiyama, N., Takamatsu, M.: NETAL: high-performance implementation of network analysis library considering computer memory hierarchy. J. Oper. Res. Soc. Jpn. **54**(4), 259280 (2011)
18. Yasui, Y., Fujisawa, K., Sato, Y.: Fast and energy-efficient breadth-first search on a single NUMA system. In: Intentional Supercomputing Conference (ISC 14) (2014)

ppOpen-HPC: Open Source Infrastructure for Development and Execution of Large-Scale Scientific Applications on Post-Peta-Scale Supercomputers with Automatic Tuning (AT)

Kengo Nakajima, Masaki Satoh, Takashi Furumura, Hiroshi Okuda, Takeshi Iwashita, Hide Sakaguchi, Takahiro Katagiri, Masaharu Matsumoto, Satoshi Ohshima, Hideyuki Jitsumoto, Takashi Arakawa, Futoshi Mori, Takeshi Kitayama, Akihiro Ida and Miki Y. Matsuo

Abstract ppOpen-HPC is an open source infrastructure for development and execution of large-scale scientific applications on post-peta-scale (pp) supercomputers with automatic tuning (AT). ppOpen-HPC focuses on parallel computers based on many-core architectures and consists of various types of libraries covering general procedures for scientific computations. The source code, developed on a PC with a single processor, is linked with these libraries, and the parallel code generated is optimized for post-peta-scale systems. In this article, recent achievements and progress of the ppOpen-HPC project are summarized.

Keywords ppOpen-HPC · Post-peta-scale systems · Automatic tuning · Parallel computing

---

K. Nakajima (✉) · M. Satoh · T. Furumura · H. Okuda · T. Katagiri · M. Matsumoto · S. Ohshima · F. Mori · T. Kitayama
The University of Tokyo, Tokyo, Japan
e-mail: nakajima@cc.u-tokyo.ac.jp

M. Satoh
e-mail: satoh@aori.u-tokyo.ac.jp

T. Furumura
e-mail: furumura@eri.u-tokyo.ac.jp

H. Okuda
e-mail: okuda@k.u-tokyo.ac.jp

T. Katagiri
e-mail: katagiri@cc.u-tokyo.ac.jp

M. Matsumoto
e-mail: matsumoto@cc.u-tokyo.ac.jp

S. Ohshima
e-mail: ohshima@cc.u-tokyo.ac.jp

K. Fujisawa et al. (eds.), *Optimization in the Real World*,
Mathematics for Industry 13, DOI 10.1007/978-4-431-55420-2_2

1 Overview of ppOpen-HPC

Today, high-end parallel computer systems are becoming larger and more complex. It is very difficult for scientists and engineers to develop efficient application codes that make use of the potential performance of these systems.

We propose an open source infrastructure for development and execution of optimized and reliable simulation codes on large-scale parallel computers. This infrastructure is named *ppOpen-HPC* [1, 2], where "pp" stands for "post-peta-scale", as shown in Fig. 1. The target post-peta-scale system is the Post T2K System, which will be installed and operated by the Joint Center for Advanced High Performance Computing (JCAHPC) [3] under collaboration between the University of Tsukuba and the University of Tokyo. The Post T2K System, which will be installed in FY 2016, is based on many-core architectures, such as the Intel MIC/Xeon Phi. Its peak performance is expected to be more than 30 PFLOPS.

ppOpen-HPC is a five-year project (FY 2011–2015) and a part of the "Development of System Software Technologies for Post-Peta-Scale High Performance Computing" funded by JST/CREST (Japan Science and Technology Agency, Core Research for Evolutional Science and Technology) [4]. ppOpen-HPC is being developed by the University of Tokyo (Information Technology Center, Atmosphere and Ocean Research Institute, Earthquake Research Institute, Graduate School of Frontier Sciences), Kyoto University, Hokkaido University, and Japan Agency for Marine-Earth Science and Technology (JAMSTEC). The expertise of the members covers

F. Mori
e-mail: f-mori@eri.u-tokyo.ac.jp

T. Kitayama
e-mail: kitayama@h.k.u-tokyo.ac.jp

T. Iwashita
Hokkaido University, Hokkaido, Japan
e-mail: iwashita@iic.hokudai.ac.jp

H. Sakaguchi · M.Y. Matsuo
JAMSTEC, Kanagawa, Japan
e-mail: sakaguchih@jamstec.go.jp

M.Y. Matsuo
e-mail: mikiy@jamstec.go.jp

H. Jitsumoto
Tokyo Institute of Technology, Tokyo, Japan
e-mail: jitumoto@gsic.titech.ac.jp

T. Arakawa
RIST, Tokyo, Japan
e-mail: arakawa@rist.jp

A. Ida
Kyoto University, Kyoto, Japan
e-mail: ida@media.kyoto-u.ac.jp

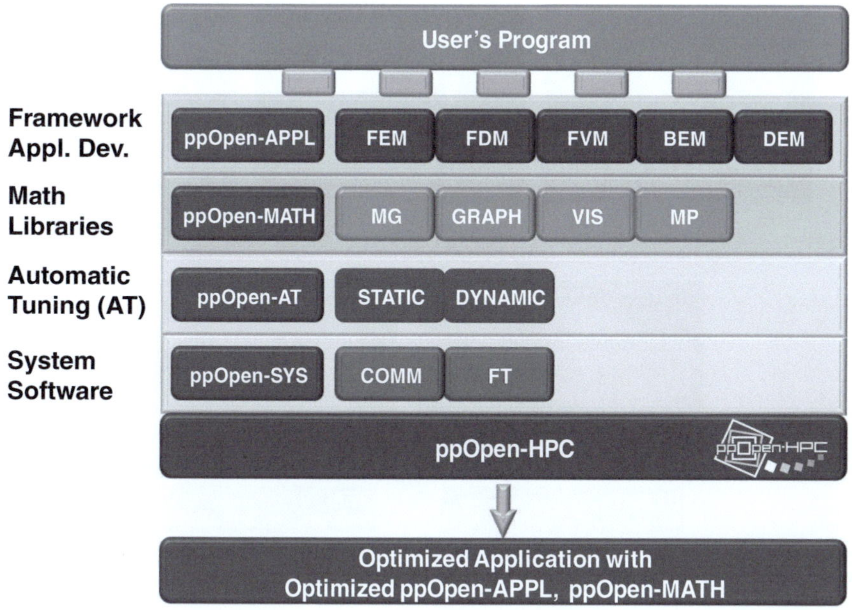

Fig. 1 Overview of ppOpen-HPC

a wide range of disciplines related to scientific computing, such as system software, numerical libraries/algorithms, computational mechanics, and earth sciences. ppOpen-HPC includes the following four components (Fig. 1):

- ppOpen-APPL
- ppOpen-MATH
- ppOpen-AT
- ppOpen-SYS

Libraries in ppOpen-APPL, ppOpen-MATH, and ppOpen-SYS are called from user programs written in Fortran and C/C++ with MPI and OpenMP.

In ppOpen-HPC, we are focusing on five types of discretization methods for scientific computing: FEM, FDM, FVM, BEM, and DEM (Fig. 2). ppOpen-APPL is a set of optimized libraries covering various types of procedures for these five methods. Source code developed on a PC with a single processor is linked with ppOpen-APPL, and the parallel code generated will be optimized for a post-peta-scale systems. Key issue for this type of framework like ppOpen-APPL is well-designed data structure for scientific computing. In previous projects, such as GeoFEM [5] and HEC-MW [6], some of the authors developed such frameworks, where typical procedures for FEM have been optimized for certain types of supercomputers. This type of framework provides dramatic efficiency, portability, and reliability in the development and execution of scientific applications. It reduces both the number of steps in the source code and the duration of time required for parallelization and

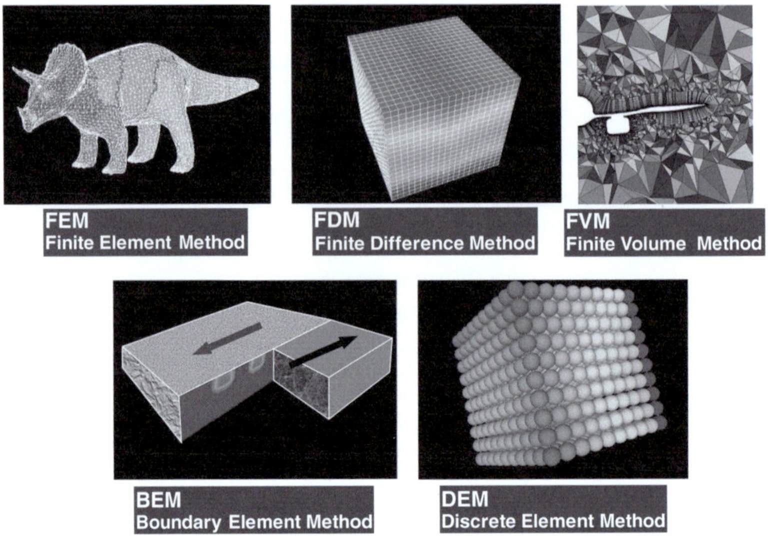

Fig. 2 Target applications of ppOpen-HPC

optimization of legacy code. In ppOpen-HPC, we extend this idea to other four types of methods, and introduce a new feature, *automatic tuning (AT)*. AT enables a smooth and easy shift to further development on future architectures through the use of ppOpen-AT, which generates optimized libraries and applications under various types of environments automatically.

ppOpen-MATH is a set of libraries for multigrid, visualization, loose coupling, etc., while ppOpen-SYS includes system software libraries related to node-to-node communication and fault tolerance. ppOpen-HPC enables more than 2,000 users of the supercomputer system in the University of Tokyo to switch from homogeneous multicore clusters to a post-peta-scale system based on many-core architectures. Although the final target of ppOpen-HPC is the Post T2K system, libraries for multi-core clusters, such as K computer, and Fujitsu PRIMEHPC FX10, are also developed.

In the following sections, we describe recent achievements in the development of each component of ppOpen-HPC, as shown in Fig. 1.

2 ppOpen-APPL

ppOpen-APPL is a set of libraries that covers various types of procedures for scientific computations, such as parallel I/O of datasets, matrix formation, linear solvers with practical and scalable preconditioners, visualization, adaptive mesh refinement (AMR), and dynamic load balancing, in various types of models, including FEM, FDM, FVM, BEM, and DEM, shown in Fig. 2.

Each component is based on existing practical application codes. ppOpen-APPL provides common data structures and interfaces that support users with the easy implementation procedures of ppOpen-HPC onto legacy codes.

2.1 Simulation of 3D Seismic Wave Propagation Using ppOpen-APPL/FDM

ppOpen-APPL/FDM is a framework for the development of applications by the finite difference method (FDM), as shown in Fig. 1.

In the recent trend of supercomputer architectures, the *byte-per-flops* ratio (B/F ratio, ratio of memory bandwidth (BYTE/sec, B) and computational performance (FLOPS, F)) has been dropping drastically. For example, it dropped from 4 to 0.5 for the Earth Simulator and the K computer. The B/F ratio of next-generation computers is expected to fall even further. Therefore, an important issue for the future high-performance parallel computing of FDM simulations is the restriction of the memory bandwidth relative to the CPU speed. To overcome this problem, it is necessary to develop a new FDM simulation structure suitable for future many-core and low B/F machines.

We proposed to effectively decrease the required B/F ratio of the FDM simulation of seismic wave propagation. To validate our proposal, we evaluated the performance of a parallel 3D FDM simulation of seismic wave propagation on the Intel Xeon Phi coprocessor [7]. The original FDM simulation code first loaded velocity and stress components from memory to processor, calculated each spatial derivative, and stored them in memory. This requires large B/F ratios to load and store the large number of variables. These derivatives were then used for the next kernels of update stress and update velocity, which also require large B/F ratios of 2.7 and 1.7 for each kernel, respectively. We modified the B/F reduction FDM code to merge the derivative and update calculations, and thereby avoided the need to store and load variables during the calculations. As a result, the required B/F ratios for both kernels dropped dramatically to 0.4.

Figure 3 compares the performance of the original and the modified code in MPI/OpenMP hybrid parallel computing on the Intel Xeon Phi processor. For parallel computing up to 60 physical cores, the modified code is slower than the original code. However, with much larger thread parallel simulation using 240 logical cores, it is confirmed that the speed of the modified code is double the speed of the original code. This B/F reduction code is also suitable for other parallel processors, such as Fujitsu PRIMEHPC FX-10 at the University of Tokyo [8].

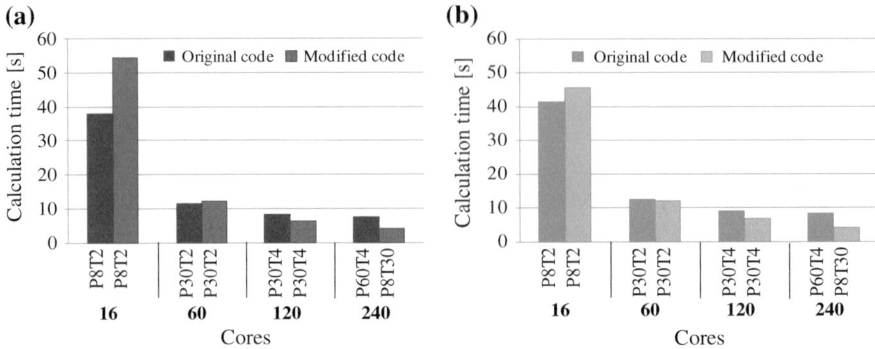

Fig. 3 Comparison of MPI/OpenMP hybrid parallel computing based on the original code and on the modified B/F reduction code for the **a** update velocity and **b** update stress kernels. Indexes *P* and *T* indicate number of MPI processes and OpenMP threads, respectively (e.g., P30T4 means hybrid parallel computing using 30 MPI processes and each has 4 OpenMP threads)

2.2 *ppOpen-APPL/AMR-FDM with Adaptive Mesh Refinement*

We developed an adaptive mesh refinement (AMR) framework for explicit FDM schemes in the ppOpen-APPL/AMR-FDM library [9]. To overcome the problem of load imbalance in parallelized AMR simulations, we implemented a dynamic domain decomposition (DDD) technique, with which the whole computational domain is dynamically re-decomposed into new subdomains so that the computational load on each process becomes nearly the same. Test simulations of a linear advection equation using the AMR framework are shown in Fig. 4. The fine grids are adaptively created where the gradient of the waveform is high.

Figure 5 shows the temporal evolutions of execution time for the simulation of 512 MPI processes in the cases with and without DDD. In the case without DDD, the profile of the execution time has large fluctuations because the waveform propagates across boundaries of the subdomains. On the other hand, in the case with DDD, the average execution time can be reduced. The DDD procedure succeeds in significantly holding down the average execution time. The graph on the right in Fig. 5 shows the execution time focused on a certain iteration count. The wave profile of the zigzag line is attributed to the timing of DDD, which takes time to perform. The DDD performs when the computational costs exceed a load-balance criterion. The time difference between DDD ON and OFF constitutes the overhead of the DDD procedure.

In addition to the above, we also developed the AMR framework for implicit time-marching schemes in which AMR and the multigrid method are used concurrently. In this implementation, each grid layer created by the AMR method corresponds to each layer of the V-cycle used by the multigrid method. We aim for not only explicit but also implicit schemes that can use the framework. Although code optimizations

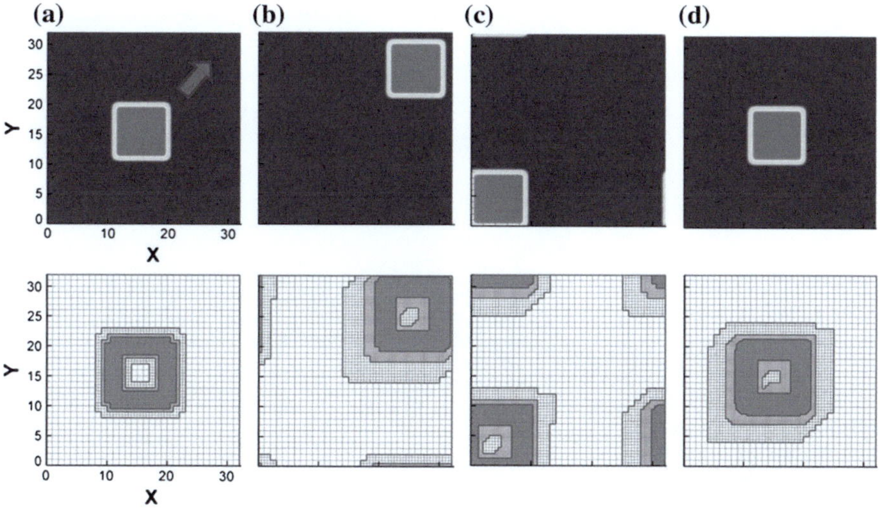

Fig. 4 Example of AMR

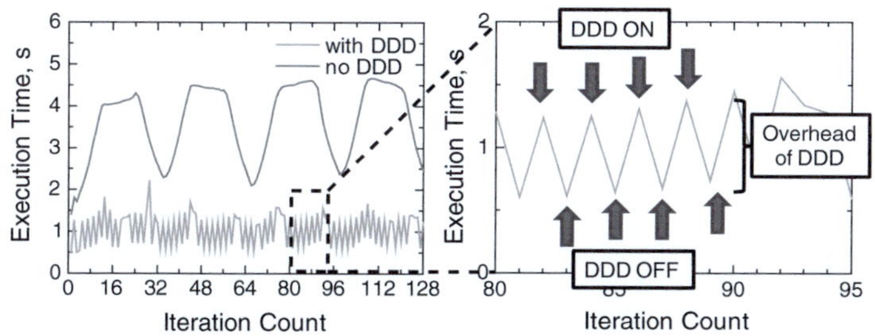

Fig. 5 Effect of DDD procedures

are needed, such an AMR framework shows the value of development for multi-scale simulations on post peta-scale systems.

2.3 ppOpen-APPL/BEM and HACApK

The ppOpen-APPL/BEM is a software tool for large-scale parallel boundary element method (BEM) analyses. This tool consists of the BEM-BB (Back-Bone) framework, templates, and the HACApK library. All the components are parallelized based on the hybrid MPI+OpenMP programming model. The BEM-BB framework provides users with parallelized program code for coefficient matrix generation and linear system solvers. Users can easily develop a parallel BEM code for their own applications by adding a user function describing the integral operation onto the framework. For some

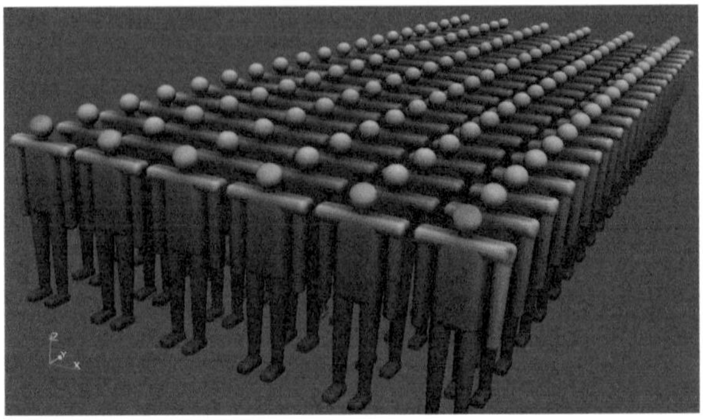

Fig. 6 Example of static electric field analysis: The electrical charge on surfaces of the humanoids is calculated

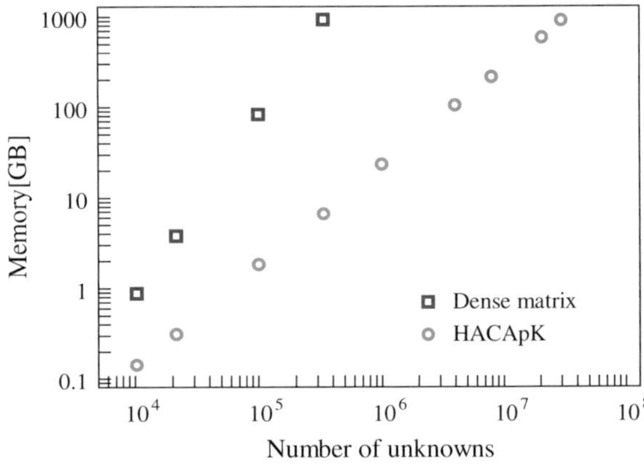

Fig. 7 Memory usage of H-matrices and original dense matrices as a function of the number of unknowns N

specific application domains, template programs for integral operation are provided. By using the BEM-BB framework with a template, the user obtains complete BEM simulation code. Currently, a template for static electric field analysis is available (Fig. 6). Moreover, we have been developing the HACApK library to realize faster and larger-scale BEM analyses.

The HACApK library adopts hierarchical matrices (H-matrices) with adaptive cross approximation (ACA) as the approximation technique for dense matrices occurring from the integral equation method represented by BEM. The method of H-matrices with ACA is based on the idea that submatrices corresponding to remote interactions become numerically low-rank matrices. H-matrices with ACA reduce the complexity from $O(N^2)$ to $O(NlogN)$, where N denotes the number of unknowns (Fig. 7).

For parallelization of H-matrices on symmetric multiprocessor (SMP) cluster systems, we proposed a set of algorithms for constructing H-matrices and performing multiplication of an H-matrix and a vector [10]. The proposed algorithms are implemented by the flat-MPI and hybrid MPI+OpenMP programming models. The performance of these implementations is evaluated by electric field analysis (Fig. 8). In the flat-MPI version, the speedup is limited in hierarchical matrix-vector multiplication.

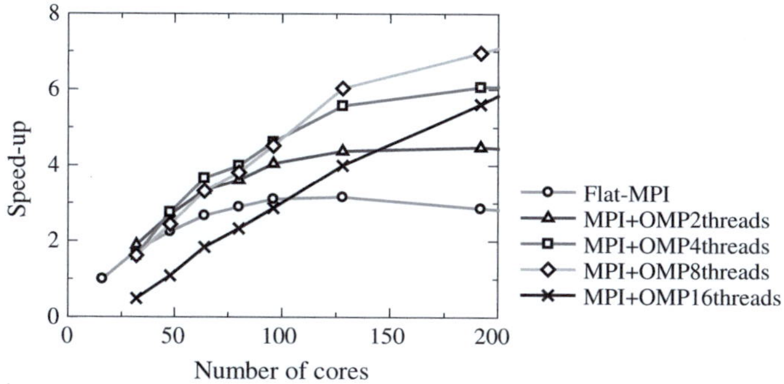

Fig. 8 Parallel scalability when performing a multiplication of an H-matrix and a vector on Fujitsu PRIMEHPC FX10 at the University of Tokyo [8]

We succeeded in developing a hybrid MPI+OpenMP version to improve the parallel scalability. In numerical experiments, the hybrid version exhibits a better parallel speed-up for the hierarchical matrix-vector multiplication up to 256 cores of Fujitsu PRIMEHPC FX10 at the University of Tokyo [8]. In addition to the above parallel algorithms, we also proposed an improved method for H-matrices with ACA [11]. By using the proposed method, we can avoid the problem that ranks of approximated matrices increase rapidly as the matrix size increases when conventional H-matrices with ACA are employed for an integral equation whose kernel function has high-order singularities. In particular, application of the proposed method enables us to perform large-scale simulations, in which conventional H-matrices with ACA fail to construct appropriate low-rank approximations.

2.4 ppOpen-APPL/DEM: Open Library for Discrete Element Method

Particle methods are among the more commonly used approaches for numerical simulations of physical problems. Particle methods, e.g., the discrete element method (DEM), are applied in various fields such as molecular bioscience, material science, civil engineering, oceanography, and astrophysics. However, millions or billions of particles are necessary to simulate physical problems with sufficient accuracy, and this level of refinement is indeed important in applications of industrial problems. Thus, further evolution of both the infrastructure and the performance development of particle simulation is needed. ppOpen-APPL/DEM is a library designed to enable the easy implementation of particle simulation code with the interactions of short-range particles such as DEM for parallel computers.

The ppOpen-APPL/DEM main library provides fundamental subroutines and functions that organize the particle simulation coded with OpenMP/MPI hybrid programing [12]. The code using the library performs parallel computation on a PC

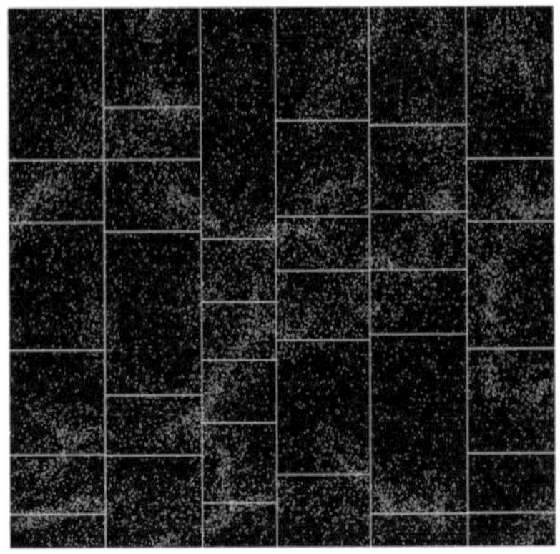

Fig. 9 Particle simulation by ppOpen-APPL/DEM: The dynamic load balance is achieved with the slice-grid method

cluster and a supercomputer. Although simulating the motions of interacting particles is easy, moving particles make it difficult to balance the computational load. To realize the dynamic load balance over the computational nodes, we implemented the DDD technique that decomposes the computational domain into several subdomains, each of which is associated with a distinct computational node (Fig. 9), which shows our implementation of the slice-grid method. Other protocols, such as orthogonal recursive bisection, will be implemented in the future. Furthermore, to reduce the difficulty of programing complex initial and boundary conditions, ppOpen-APPL/DEM provides a utility library named ppOpen-APPL/DEM-Util. This utility includes subroutines that combine a stereolithography (STL) data file into particle simulations. For example, input STL data are used to design the initial coordination of particles; the initial coordination enables the user to easily prepare a complex configuration of particle positions (Fig. 10). As another example, the input STL data are used as a boundary condition of the particle simulation. Those implementations are useful for the design of large-scale simulations.

Fig. 10 Particle simulation using ppOpen-APPL/DEM-Util: The utility enables us to use stereolithography (STL) data as the initial condition

3 ppOpen-MATH

ppOpen-MATH consists of common numerical libraries, such as multigrid solvers (ppOpen-MATH/MG) (Fig. 4), parallel graph libraries (ppOpen-MATH/GRAPH), parallel visualization (ppOpen-MATH/VIS), and a library for coupled multi-physics simulations (ppOpen-MATH/MP).

3.1 ppOpen-MATH/MP and NICAM-COCO Coupling

ppOpen-MATH/MP is a coupling software applicable to the models employing various discretization methods such as FDM, finite volume method (FVM) and finite element method (FEM) [13]. To demonstrate the applicability of ppOpen-MATH/MP, we used it for an atmospheric model and ocean model coupling. The atmospheric model selected for this purpose is the Nonhydrostatic ICosahedral Atmospheric Model (NICAM), which is a nonhydrostatic global model employing an icosahedral grid system and an FVM discretization method [14]. The CCSR Ocean Component Model (COCO) is used as an ocean model coupled with NICAM. COCO adopts a tri-polar grid, in which the northern polar region grid points do not follow a latitude-longitude grid, and the discretization method is FDM [15]. For realizing wide applicability, ppOpen-MATH/MP is designed so that users can implement their own interpolation code. The interpolation code is based on the first-order conservative remapping scheme in [16]. Physical quantities exchanged from NICAM to COCO are 13 variables, including wind speed, heat flux, and precipitation. The quantities exchanged from COCO to NICAM are 6 variables, including SST and sea ice thickness.

In addition to NICAM-COCO coupling, we implemented NICAM and IO component coupling. The reason for this coupling is that the icosahedral grid employed by NICAM is not suitable for analyzing the results. For example, the calculation of zonal mean values is not straightforward and the visualization tools assume a latitude-longitude grid (lat-lon grid) in many cases. So, we developed an IO program that converts the icosahedral grid to the lat-lon grid and is executed in parallel with NICAM. The implemented conversion schemes are a bilinear interpolation, a control volume weighted average, and the nearest-neighbor method.

Figure 11 is a schematic of the coupling system described above. The coupling system is designed so that NICAM automatically detects the coupling pattern at runtime without any other configuration. For example, when COCO is executed in parallel with NICAM, subroutines for NICAM-COCO coupling are used, and if not, subroutines of the mixed layer ocean model are called. IO is also the same as the case of COCO, in which NICAM automatically sends output data to IO only when the IO component is executed.

For utilizing ppOpen-MATH/MP, a correspondence table of a grid point index between the models is required as input data. Therefore, we developed a calculation

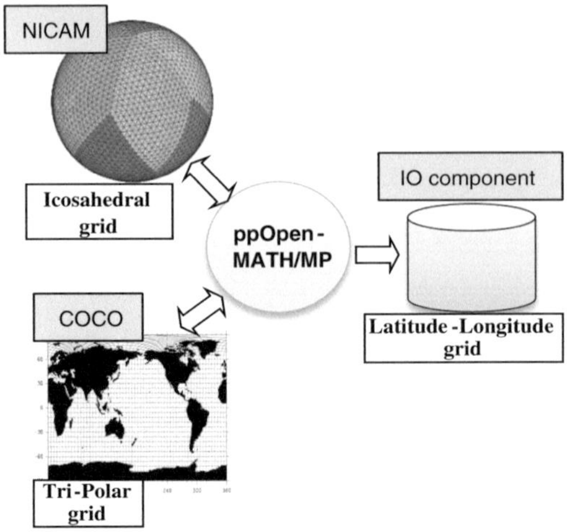

Fig. 11 Schematic of the coupling system

tool named ppOpen-MATH/MP-PP, which targets two-dimensional meshes on a sphere surface. For calculating the correspondence, a new search algorithm, for which efficiency $O(n)$ was developed, requires $O(n^2)$ calculation through the brute-force method. Figure 12 shows the result of the performance measurement of the new algorithm. Here, two mesh types, the NICAM icosahedral grid and the I/O lat-lon grid, were selected as a test case. The numbers of grid points are listed in Table 1. As shown in the figure, execution time versus the number of grid points increases at a rate of $O(n)$, and the effectiveness of the new algorithm is confirmed.

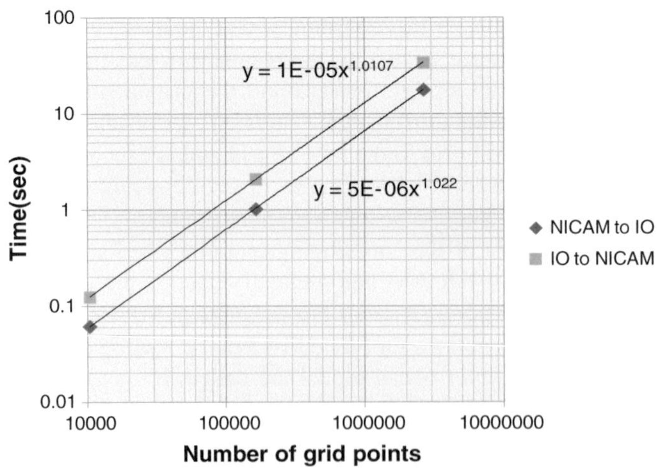

Fig. 12 Search time on NICAM grid versus lat-lon grid

Table 1 The number of grid points

The number of Lat-Lon grid	10,585	166,753	2,657,665
The number of NICAM grid	20,480	327,680	5,242,880

3.2 Integrated Earthquake Simulations Using ppOpen-MATH/MP

Simulations including multi-scale or varied physical phenomena are difficult because such kinds of modeling and implementation by a specific application are difficult. For example, to simulate the earthquake shock coming from earthquake sources and building damage, both a seismic wave that propagates over a wide region several hundred kilometers square and a shaking building that occurs in a small region of several tens of meters square must be resolved concurrently. In this case, if the FDM application using ppOpen-APPL/FDM is suitable for the analysis of elastic (seismic) wave propagations and the FEM application using ppOpen-APPL/FEM is suitable for the analysis of dynamic solid mechanics (the building) can be used in combination, the multi-scale and multi-physics coupling simulation is easy to use. Our final goal is to develop the application coupler (ppOpen-MATH/MP) [11] by using ppOpen-APPL libraries in various combinations. Figure 13 shows an example of a multi-scale and multi-physics coupling simulation by using ppOpen-MATH/MP. In this case, the seismic wave calculated by the FDM model is transferred to the FEM model and is interpolated in the FEM model mesh through ppOpen-MATH/MP. The number of nodes and processes in the FDM model and FEM model, respectively, are different, but they are automatically interpolated and are arranged by the coupler. Application

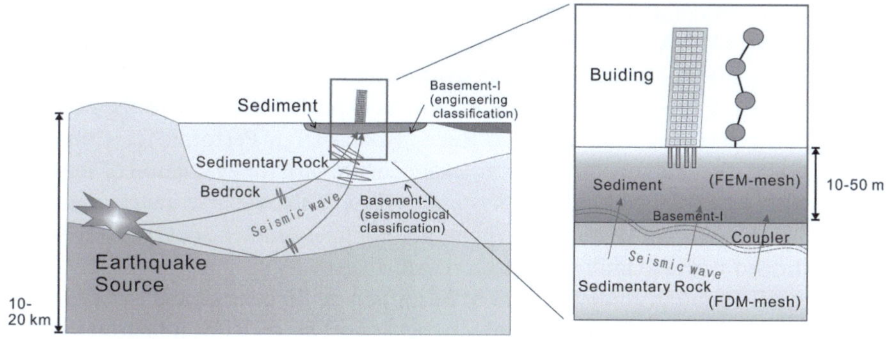

Fig. 13 Example of a multi-scale and multi-physics coupling simulation: seismic wave propagation (ppOpen-APPL/FDM), dynamic solid mechanics (ppOpen-APPL/FEM), and application coupler (ppOpen-MATH/MP)

developers need not consider the structure of the FDM and FEM applications, and they can develop a coupling simulation by using ppOpen-MATH/MP.

A practical simulation is executed on the large-scale computational resources of Fujitsu PRIMEHPC FX10 at the University of Tokyo [8]. The simulation target is the earthquake that occurred at Awaji Island on 13 April 2013. The seismic source was located on the central part of Awaji Island, Hyogo prefecture, Japan. The computational domain of Seism3D+, which is composed by the ppOpen-APPL/FDM library, is 60 km square from Awaji Island and the domain of FrontISTR++, which is composed by the ppOpen-APPL/FEM library, is an actual building of the RIKEN Advanced Institute for Computational Science (AICS), Port Island, Kobe, as modeled by an unstructured mesh. This building mesh is placed on two locations of Port Island and the Kobe stadium, where ground conditions differ from each other. The total computational nodes on the FX10 used in the simulation are 4560 nodes (16 cores/1 node): 2560 nodes for Seism3D+ and 1000 nodes/1 place = total 2000 nodes for FrontISTR++. In the simulation, seismic wave propagations (Seism3D+) for the simulation time of 90 s were calculated in the computational time of about 6 h, and building vibrations originated from the seismic wave (FrontISTR++) for the simulation time of 20 s were calculated in the computational time of about 16 h. However, it was revealed that the way to allocate memory of the coupler has some problems when such a large-scale simulation is performed. This is because a part of the initialization routine in ppOpen-MATH/MP includes the centralized procedure of all MPI processes. Optimization of the code will be carried out as a future plan.

3.3 ppOpen-MATH/MG: Multigrid Solver

Optimization of both serial and parallel communications is a critical issue for the development of scalable algorithms in next-generation applications. Serial communication is the data transfer through memory hierarchies of each processor, whereas parallel communication is the message passing between computing nodes through the network by MPI. A multigrid is a scalable method for solving linear equations and preconditioning Krylov iterative linear solvers, and is especially suitable for large-scale problems. The parallel multigrid method is expected to be one of the powerful tools on post-peta/exa-scale systems. Recently, the High Performance Conjugate Gradient (HPCG) [17] was proposed as a new benchmark for evaluation of the practical performance of supercomputer systems. HPCG solves sparse matrices derived from finite element applications by using a conjugate gradient (CG) linear solver preconditioned with the multigrid method (MGCG).

The parallel multigrid method and MGCG include both serial and parallel communication processes that are generally expensive. This article summarizes recent efforts of the optimization of serial and parallel communications in parallel MGCG solvers with geometric multigrid procedures using up to 4,096 nodes (65,536 cores) of the Fujitsu PRIMEHPC FX10 at the University of Tokyo [8]. The target application, pGW3D-FVM, is a 3D finite-volume simulation code, which solves groundwater

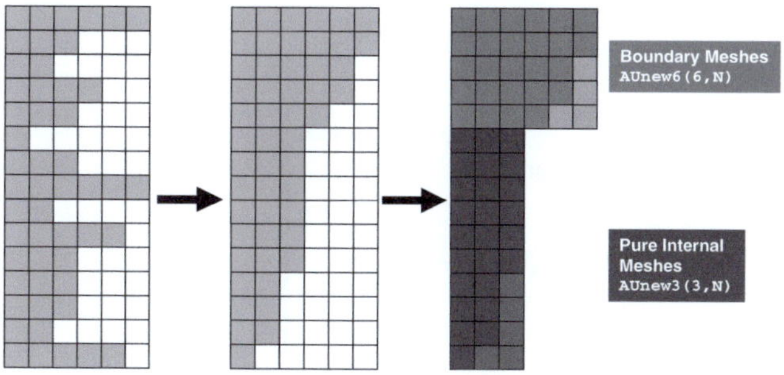

Fig. 14 Idea of sliced ELL format [19] with two slices in the present work

flow problems through heterogeneous porous media by the parallel MGCG method [18]. The performance of both flat MPI and OpenMP/MPI hybrid parallel programming model (HB M×N (*M*: number of threads on each MPI process, *N*: number of MPI processes on each node)) has been evaluated.

In the present work, a new format for sparse matrix storage based on sliced ELL [19] (Fig. 14), which has been well utilized for the optimization of sparse matrix-vector multiplication (SpMV), is proposed for optimization of serial communication on memories. In addition, hierarchical coarse grid aggregation (*h*CGA) (Fig. 15) is introduced for optimization of parallel communication by message passing. The proposed methods are implemented for pGW3D-FVM, and the robustness and performance of the code was evaluated by using up to 4,096 nodes (65,536 cores) of the Fujistu FX10 system. The parallel MGCG solver using the *sliced* ELL format

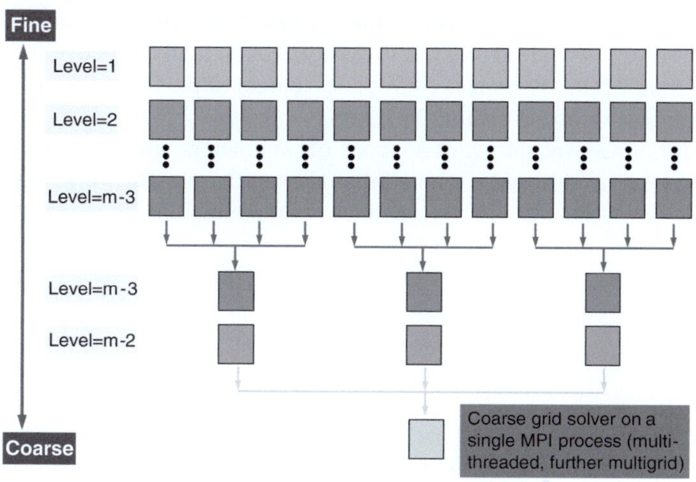

Fig. 15 Procedures of hierarchical CGA (*h*CGA), where the number of MPI processes is reduced before the final coarse grid solver of CGA on a single MPI process

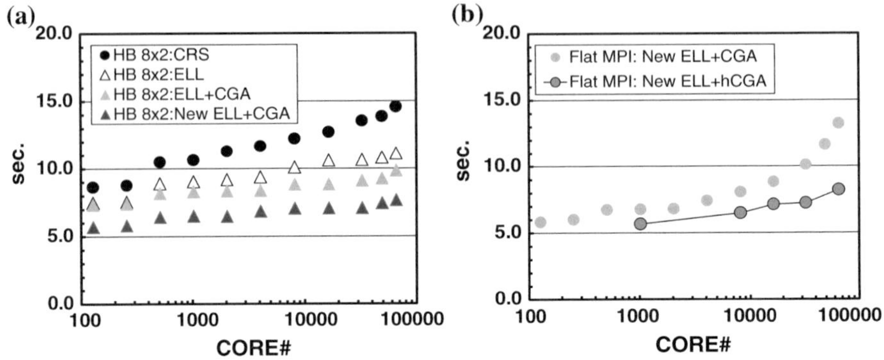

Fig. 16 Performance of MGCG solver on Fujitsu FX10 using up to 4,096 nodes (65,536 cores), weak scaling (elapsed time for MGCG): 262,144 ($=64^3$) meshes/core, max. total problem size: 17,179,869,184 meshes

provided performance improvement in both weak scaling (25–31 %) and strong scaling (9–22 %) compared to the code using the original ELL format. Moreover, *h*CGA provided excellent performance improvement in both weak scaling (1.61 times) and strong scaling (6.27 times) for the flat MPI parallel programming model. *h*CGA was also effective for improvement of parallel communications (Fig. 16a, b).

The effect of sliced ELL on serial communication was significant, while that of *h*CGA on parallel communication was not so impressive except for flat MPI cases. Because *h*CGA proved to be very effective for reducing the overhead of the coarse grid solver, it will also provide a more significant effect on hybrid parallel programming models with a larger number of nodes. The computational amount of the coarse grid solver for each core of flat MPI is 256 ($=16 \times 16$) times as large as that of HB 16×1. Therefore, *h*CGA is expected to be really effective for HB 16×1 with more than 2.50×10^5 nodes (4.00×10^6 cores) of the Fujitsu FX10, where the peak performance is more than 60 PFLOPS.

CGA and *h*CGA include various types of parameters, and the optimum values of those were derived through empirical studies in the present work. Development of methods for automatic selection of these parameters [20] is also an interesting technical issue for future work.

4 ppOpen-AT

Computer architectures are becoming more and more complex due to non-standardized memory accesses and hierarchical caches. The AT capability is an important and critical technology for further development of new architectures, and maintenance of the overall framework to establish high productivity for performance tuning. Generally speaking, keeping high performance with a legacy code

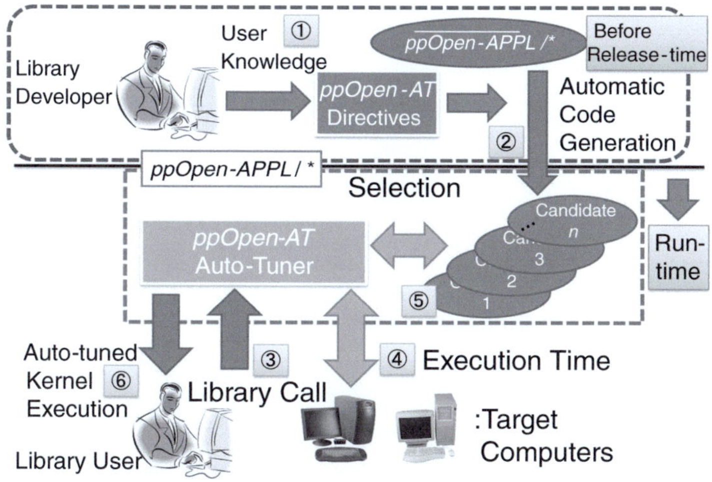

Fig. 17 Procedures for generation of optimized code by using ppOpen-AT

on different computer environments, described as "performance portability", is a challenging issue. In this section, we show the current results of research on the AT function to establish performance portability.

ppOpen-AT automatically and adaptively generates optimum implementations for efficient memory accesses in the processes of methods for scientific computing in each component of ppOpen-APPL. Example processes are explicit time marching procedures, matrix assembling procedures, and implicit linear solvers. These are achieved under various environmental constraints, such as the architecture of the supercomputer system, available resources, problem size, etc. ppOpen-AT also optimizes widely used open source applications and numerical libraries, such as OpenFOAM and PETSc. With the focus on optimum memory access, directive-based special AT languages for specific procedures in scientific computing are being developed.

Figure 17 describes procedures for the generation of optimized code using ppOpen-AT.

We applied ppOpen-AT to simulation code based on FDM, which was provided as ppOpen-APPL/FDM in ppOpen-HPC. The framework utilizes well-known loop transformation techniques, such as loop fusion and loop split. In addition, we looked at the AT function to support compiler optimizations with a re-ordering of statements. Processes of data packing and unpacking for communication with MPI are targets for AT. The components of AT are carefully designed to minimize the use of software stacks to satisfy the requirements of the many-core architectures currently in operation. The FIBER framework [19] is utilized to implement an auto-tuner and timings of invocation of the AT function. In particular, execution of the AT with dedicated problem sizes and numbers of threads is crucial. The timing called Before execute-time AT on FIBER is used to optimize the target codes of ppOpen-HPC.

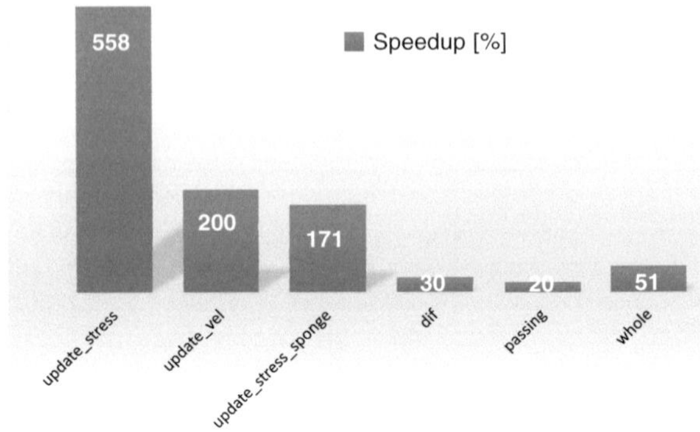

Fig. 18 Effect of AT by ppOpen-AT on the performance of ppOpen-APPL/FDM

The results of evaluations conducted using ppOpen-AT indicate that maximum speedup factors greater than 550 % are obtained when ppOpen-AT is applied in eight nodes of the Intel Xeon Phi [21] (Fig. 18). To show performance portability, an additional evaluation with the Intel Ivy Bridge and the Sparc64 IX-fx was performed. Different parameters were set by the AT. For execution with auto-tuned parameters, the Intel Xeon Phi is faster than the other two architectures, while the Xeon Phi is the slowest if we do not apply the AT. This shows that AT is a crucial factor to establish performance portability.

5 ppOpen-SYS/FT: Application-Level Checkpoint/Restart (CP/RS) Framework with Runtime Optimization

The application-level checkpoint/restart technique is periodically makes snapshots of application and stores them for recovery later by application programming. And it is frequently implemented within the application which has time stepping. However, optimizations of some parameters require the runtime information. In this case, the parameters described in the program source tend to depend on the application programmer's ad hoc decision.

For reducing the cost of checkpoint/restart (CP/RS) at the application level, the framework focused on a time-stepping model application and asynchronous coordination on the checkpoint with an applied optimized parameter. The scenario is as follows: (1) Processes send runtime information to the optimizing daemon. (2) The daemon optimizes the runtime with information of the processes and the environment, and returns the result to the processes. (3) The processes make a checkpoint with the newly optimized information. (4) The daemon deletes a waste checkpoint with appropriate information, such as the core-loop stepping count included on the checkpoint ppOpen-SYS/FT framework that has a reference implementation for checkpoint interval optimization.

ppOpen-SYS/FT is the directive-based application-level checkpoint/restart framework and its implementation with runtime optimization [22–24]. The implementation of ppOpen-SYS/FT adapts to large-scale systems managed with a job scheduler. This implementation optimizes the interval with the count of stepping, the execution time of stepping, and checkpointing. The optimization daemon returns the appropriate interval as an offset or stride format. Then, if the process passes the offset, it ignores the offset and the checkpoint on the next timing. This implementation is used with the directive-based method. In addition, we are considering the partial message logging (PML) method by the ppOpen-SYS/FT framework. This method partitions application processes as a group and uses a different checkpoint method on the intra-group and the inter-group. The partitioning depends on the communication amount and the frequency between processes. ppOpen-SYS/FT optimizes this partitioning with the topological method that we also proposed (Fig. 19).

Moreover, we are considering cooperation with ppOpen-MATH/MP. By focusing communication between the coupler and applications that construct coupling software, each application can perform checkpointing individually with our asynchronous coordination. However, because of the coupler keeping state, the consistency of the coupling software is broken. Consequently, ppOpen-SYS/FT supplies an API that informs whether the application will be a checkpoint until the next communication with the coupler. On each communication of coupled applications, the coupler calls the API for the decision of the checkpoint itself (Fig. 20). An additional API is also defined for AMR applications with ppOpen-APPL/FVM. The AMR application changes the checkpoint time drastically by re-partitioning. Then, ppOpen-SYS/FT supplies the optimized checkpoint timing and receives the checkpoint timing from the application decision.

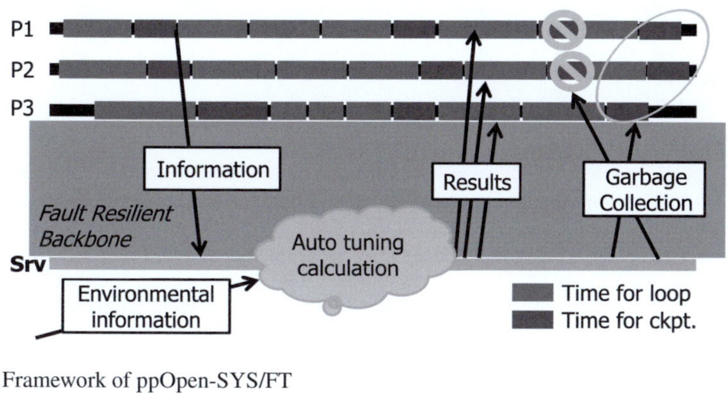

Fig. 19 Framework of ppOpen-SYS/FT

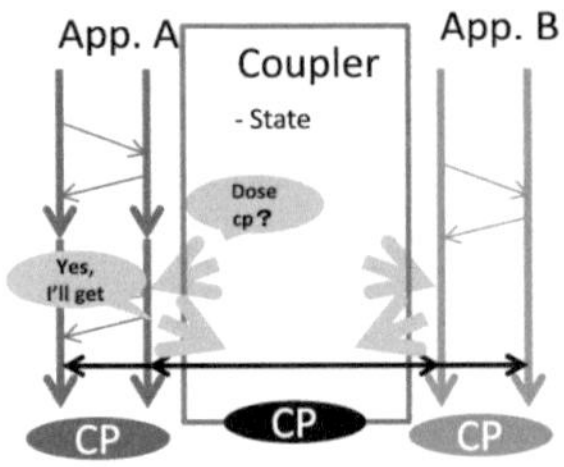


```
appli_A(){
  for(i){
    calc_A();
    if(i%x == 0)
      coupsend();
      cp_cand();
  }
}
```

```
coupsend(){
  data_renew();
  if(checkCP()){
    cp();
  }
}
```

```
appli_B(){
  for(i){
    calc_B();
    if(i%y == 0)
      couprecv();
      cp_cand();
  }
}
```

Fig. 20 Procedures for checkpointing of couplers

6 Summary

In this article, recent achievements and progress of the ppOpen-HPC project have been presented. The libraries developed for ppOpen-HPC are open for public use under MIT license and can be downloaded at the website of the project [1]. ppOpen-HPC has been installed on various types of supercomputers, and is utilized for research and development that requires large-scale supercomputer systems. Moreover, ppOpen-HPC is introduced in graduate and undergraduate classes at universities.

Currently, we are focusing on development and optimization of ppOpen-HPC for Intel Xeon/Phi architecture, and preparing for further research and development towards exascale systems.

Acknowledgments This work is supported by Core Research for Evolutional Science and Technology (CREST), the Japan Science and Technology Agency (JST), Japan.

References

1. ppOpen-HPC: http://ppopenhpc.cc.u-tokyo.ac.jp/
2. Nakajima, K.: ppOpen-HPC: open source infrastructure for development and execution of large-scale scientific applications on post-peta-scale supercomputers with automatic tuning (AT). In: ATIP '12 Proceedings of the ATIP/A*CRC Workshop on Accelerator Technologies for High-Performance Computing: Does Asia Lead the Way?, ACM Digital Library (ISBN: 978-1-4503-1644-6) (2012)
3. Joint Center for Advanced High Performance Computing (JCAHPC): http://jcahpc.jp/
4. Post-Peta CREST: http://postpeta.jst.go.jp/en/
5. GeoFEM: http://geofem.tokyo.rist.or.jp/
6. HEC-MW: http://www.multi.k.u-tokyo.ac.jp/FrontISTR/
7. Mori, F., Matsumoto, M., Furumura, T.: Performance optimization of the 3D FDM simulation of seismic wave propagation on the intel Xeon Phi coprocessor using the ppOpen-APPL/FDM library. In: Lecture Notes in Computer Science (LNCS) (in press)
8. Information Technology Center, The University of Tokyo: http://www.cc.u-tokyo.ac.jp
9. Matsumoto, M., Mori, F., Ohshima, S., Jitsumoto, H., Katagiri, T., Nakajima, K.: Implementation and evaluation of an AMR framework for FDM applications. Procedia Comput. Sci. **29**, 936–946 (2014)

10. Ida, A., Iwashita, T., Mifune, T., Takahashi, Y.: Parallel hierarchical matrices with adaptive cross approximation on symmetric multiprocessing clusters. J. Inf. Process. **22**(4), 642–650 (2014)
11. Ida, A., Iwashita, T., Ohtani, M., Hirahara, K.: Improvement of hierarchical matrices with adaptive cross approximation for large-scale simulation. IPSJ Trans. Adv. Comput. Syst. **49** (in press)
12. Nishiura, D., Matsuo, M.Y., Sakaguchi, H.: ppohDEM: computational performance for open source code of the discrete element method. Comput. Phys. Commun. **185**, 1486–1495 (2014)
13. Arakawa, T., Inoue, T., Satoh, M.: Performance evaluation and case study of a coupling software Ppopen-MATH/MP. Procedia Comput. Sci. **29**, 924–935 (2014)
14. Satoh, M., Tomita, H., Yashiro, H., Miura, H., Kodama, C., Seiki, T., Noda, A.T., Yamada, Y., Goto, D., Sawada, M., Miyoshi, T., Niwa, Y., Hara, M., Ohno, T., Iga, S., Arakawa, T., Inoue, T., Kubokawa, H.: The non-hydrostatic icosahedral atmospheric model: description and development. In: Progress in Earth and Planetary Science, pp. 1–18 (2014)
15. Hasumi, H.: Documentaion for CCSR Ocean Component Model (COCO) Version 4.0s. Center for Climate System Research, April (2007)
16. Jones, P.H.: First- and second-order conservative remapping schemes for grids in spherical coordi-nates. Mon. Weather Rev. **127**, 2204–2210 (1999)
17. HPCG: High Performance Conjugate Gradients: https://software.sandia.gov/hpcg/
18. Nakajima, K.: Optimization of serial and parallel communications for parallel geometric multigrid method. In: Proceedings of the 20th IEEE International Conference for Parallel and Distributed Systems (ICPADS 2014), pp. 25–32 (2014)
19. Monakov, A., Lokhmotov, A., Avetisyan, A.: Automatically tuning sparse matrix-vector multiplication for GPU architectures. Lect. Notes Comput. Sci. **5952**, 112–125 (2010)
20. Nakajima, K.: Automatic tuning of parallel multigrid solvers using OpenMP/MPI hybrid parallel programming models. Lect. Notes Comput. Sci. **7851**, 435–450 (2013)
21. Katagiri, T., Ohshima, S., Matsumoto, M.: Auto-tuning of computation kernels from an FDM Code with ppOpen-AT. In: Proceedings of IEEE MCSoC2014, pp. 91–98 (2014). doi:10.1109/MCSoC.2014.22
22. Jitsumoto, H., Todoroki, Y., Ishikawa, Y., Sato, M.: Grid-oriented process clustering system for partial message logging. In: Proceedings of the 4th Fault Tolerance for HPC at eXtreme Scale (FTXS) 2014, in conjunction with DSN2014 (2014)
23. Jitsumoto, H., Todoroki, Y., Sato, M.: Design and evaluations of application based fault tolerance framework with stencil model. In: G8 ESC Workshop at Kobe (2014)
24. Jitsumoto, H., Kamoshida, Y.: Application-level checkpoint/restart framework with optimal checkpoint interval. In: HPC in Asia Workshop Poster Session at ISC'13 (2013)
25. Katagiri, T., Kise, K., Honda, H., Yuba, T.: FIBER: a general framework for auto-tuning software. Proc. ISHPC-V, Lect. Notes Comput. Sci. **2858**, 146–159 (2003)

Structure-Based Primal Heuristics for Mixed Integer Programming

Gerald Gamrath, Timo Berthold, Stefan Heinz and Michael Winkler

Abstract Primal heuristics play an important role in the solving of mixed integer programs (MIPs). They help to reach optimality faster and provide good feasible solutions early in the solving process. In this paper, we present two new primal heuristics which take into account global structures available within MIP solvers to construct feasible solutions at the beginning of the solving process. These heuristics follow a large neighborhood search (LNS) approach and use global structures to define a neighborhood that is with high probability significantly easier to process while (hopefully) still containing good feasible solutions. The definition of the neighborhood is done by iteratively fixing variables and propagating these fixings. Thereby, fixings are determined based on the predicted impact they have on the subsequent domain propagation. The neighborhood is solved as a sub-MIP and solutions are transferred back to the original problem. Our computational experiments on standard MIP test sets show that the proposed heuristics find solutions for about every third instance and therewith help to improve the average solving time.

Keywords Large neighborhood search · Mixed-integer programming · Primal heuristics · Domain propagation

G. Gamrath (✉)
Zuse Institute Berlin, Takustr. 7, 14195 Berlin, Germany
e-mail: gamrath@zib.de

T. Berthold · S. Heinz
Fair Isaac Europe Ltd, c/o ZIB, Takustr. 7, 14195 Berlin, Germany
e-mail: timoberthold@fico.com

S. Heinz
e-mail: stefanheinz@fico.com

M. Winkler
Gurobi GmbH, c/o ZIB, Takustr. 7, 14195 Berlin, Germany
e-mail: winkler@gurobi.com

K. Fujisawa et al. (eds.), *Optimization in the Real World*,
Mathematics for Industry 13, DOI 10.1007/978-4-431-55420-2_3

1 Introduction

Mixed integer linear programming problems (MIPs) minimize (or maximize) a linear objective function subject to linear constraints and integrality restrictions on a part of the variables. More formally, a MIP is stated as follows:

$$z_{MIP} = \min\{c^T x : Ax \leq b, \ell \leq x \leq u, x_i \in \mathbb{Z} \text{ for all } i \in \mathcal{I}\} \tag{1}$$

with objective function $c \in \mathbb{R}^n$, constraint matrix $A \in \mathbb{R}^{m \times n}$, and constraint right-hand sides $b \in \mathbb{R}^m$. We allow lower and upper bounds $\ell, u \in \bar{\mathbb{R}}^n$ on variables, where $\bar{\mathbb{R}} := \mathbb{R} \cup \{\pm\infty\}$, and the restriction of a subset of variables $\mathcal{I} \subseteq \mathcal{N} = \{1, \ldots, n\}$ to integral values. In the remainder of this paper, we denote by $\mathcal{P}(c, A, b, \ell, u, \mathcal{I})$ a MIP of form (1) in dependence of the provided data.

This allows to model many real-world optimization problems from various fields like production planning [28], scheduling [20], transportation [13], or telecommunication networks [24]. On the other hand, the strict specifications for the problem statement make it possible to solve arising optimization problems for all these applications using the same algorithm. Therefore, very powerful generic solvers for MIPs have been developed over the last decades, which are used widely in research and practice [5, 11, 25].

These solvers are based on a branch-and-bound algorithm [15, 23], which is accalerated by various extensions. The basic concept is to split the problem into subproblems, until they are easy enough to be solved. During this process, a lower bound is computed for each subproblem by solving its linear programming (LP) relaxation $\mathcal{P}(c, A, b, \ell, u, \emptyset)$, that is the problem obtained from (1) by omitting the integrality restrictions. At the same time, the objective value of the incumbent—the best feasible solution found so far—provides an upper bound on the global optimum. In combination, these bounds allow to speed up the solving process by disregarding subproblems whose lower bound exceeds the upper bound since those cannot lead to an improving solution.

It is evident that this algorithm profits directly from finding good solutions as early as possible. On the one hand, these solutions originate from integral LP relaxation solutions, on the other hand, so-called *primal heuristics* try to construct new feasible solutions or improve existing ones. Primal heuristics are incomplete methods without any success or quality guarantee which nevertheless are beneficial on average. There are different common approaches applied by many heuristics, e.g., rounding of the LP solution or diving, which iteratively changes the current subproblem temporarily and solves the corresponding LP relaxation until an integral solution is obtained. For more details on primal heuristics, we refer to [6, 8, 18]. In this paper, we introduce two novel heuristics based on the large neighborhood search (LNS) paradigm. This concept defines a subproblem, the neighborhood, by adding restrictions to the problem, and then solves this subproblem as a MIP. A more detailed discussion of LNS is given in Sect. 2.

By modeling a specific problem as a MIP and solving it with a MIP solver, one profits from the decades of developments within this area. However, knowledge about the structure of the problem which could be exploited by a problem specific approach can hardly be fed into a MIP solver due to the generality of the algorithm. MIP solvers try to partially compensate this by detecting some common structures within the problem and exploiting them in the solving process (see [4, 31]). This detection is often done in the presolving phase, which is a preprocessing step trying to remove redundancies from the model and tighten the fomulation. Additionally, several global structures are detected in this phase and stored for later use. An overview of different global structures in MIP solvers and details about two of them, the clique table and the variable bound graph, are given in Sect. 3.

The heuristics presented in this paper define a neighborhood based on the clique table and the variable bound graph. They repeatedly fix variables and perform domain propagation to consider the direct consequences of these fixings on the domains of other variables. While this is a known approach in MIP heuristics (see, e.g., the shift-and-propagate heuristic [10]), our new heuristics take a step further and make domain propagation their driving force. They use the global structures to predict the effects of domain propagation in the fixing phase and by this determine the fixing order and fixing values for the variables. After the problem was reduced sufficiently, the remaining problem is then solved as a LNS sub-MIP.

A detailed description of the general scheme of the heuristics and how the global structures are used exactly is discussed in Sect. 4. The impact of the heuristics is evaluated by the computational experiments presented in Sect. 5. Finally, Sect. 6 gives our conclusions and an outlook.

2 Large Neighborhood Search for MIP

Large neighborhood search (LNS) heuristics are an important component of modern MIP solvers, see, e.g., [6, 8, 18, 26]. The main idea of LNS is to restrict the search for "good" solutions to a neighborhood centered at a particular reference point—typically the incumbent or another feasible solution. The hope is that such a restricted search space makes the subproblem much easier to solve, while still providing solutions of high quality. Of course, these restricted subproblems do not have to be solved to optimality; they are mainly searched for an improving solution.

To define the neighborhood, the feasible region of the MIP is restricted by additional constraints: most often variable fixings or some very restrictive cardinality constraint. A good definition of the neighborhood is the crucial point for LNS heuristics. There are different characteristics of a "good" neighborhood:

1. it should contain high quality solutions,
2. these solutions should be easy to find, and
3. the neighborhood should be easy to process.

Naturally, these three goals are conflicting in practice.

The relaxation induced neighborhood search (RINS) [16] uses two reference points: The incumbent MIP solution which fulfills the first two requirements and the optimum of the LP relaxation which fulfills the latter two. RINS defines the neighborhood by fixing all integer variables which take the same value in both solutions. In contrast to RINS, the *relaxation enforced neighborhood search* (RENS) [9] does not require an incumbent solution. RENS fixes all integer variables that take an integral value in the optimal solution of the LP relaxation. *Crossover* [6] is an improvement heuristic that is inspired by genetic algorithms [30] and requires more than one feasible solution. For a set of feasible solutions, it fixes variables that take identical values in all of them.

Local Branching [17] measures the distance to the reference point in Manhattan norm on the integer variables and only considers solutions which are inside a k-neighborhood of the reference point, where k is typically between 10 and 20. *DINS* [19] combines the ideas of RINS and Local Branching. It defines the neighborhood by introducing a distance function between the incumbent solution and the optimum of the LP relaxation. When applied during a branch-and-bound search, it further takes into account how variables change their values at different nodes of the tree.

DINS, RINS, RENS, and Crossover define their neighborhoods by variable fixings. LNS heuristics that are based on variable fixings suffer from an inherent conflict: the original search space should be significantly reduced; thus, it seems desirable to fix a large number of variables. At the same time, the more variables get fixed, the higher is the chance that the subproblem does not contain any improving solution or even becomes infeasible.

The present paper addresses this issue by the use of global structures and propagation for defining a set of variables to be fixed and an order of fixing them.

3 Global Structures in MIP Solvers

Mixed integer programs are restricted to linear constraints, a linear objective, and integrality conditions. This makes MIP solvers easily accessible and exchangeable if a MIP model is at hand. From the modeling point of view, however, there is hardly any possibility to pass additional structural information to a solver, e.g., that and how certain model variables are connected via the combinatorics of a network structure. Nevertheless, modern MIP solvers aim at detecting structures within a model and making use of them for heuristics, cutting plane separation or presolving, see e.g., [4, 31].

Examples of global structures that are detected in presolving or during root node processing include the clique table, the implication graph, the variable bound graph, multi-commodity flow structures, permutation structures, and symmetries. Multi-commodity flows and permutations are examples of rather specific constructs that occur in only a handful of models—but are crucial for solving them. Cliques and variable bound constraints, in contrast, can be found in many MIPs of different types. So far, they have been mainly used for cutting plane generation and domain propagation, see, e.g., [1]. The remainder of the section explains the clique table and the variable bound graph in more detail.

3.1 The Clique Table

A *clique* is a set of binary variables of which at most one variable can be set to one. A clique can be given directly as a linear inequality $\sum x_i \leq 1$ or derived from more general constraints such as knapsacks: given a constraint $\sum_{i \in J} w_i x_i \leq c$, each subset $J' \subseteq J$ for which $w_j + w_k > c$ for all $(j, k) \in J' \times J'$ defines a clique. In addition, presolving techniques such as probing [32] can be used to detect cliques which are only given implicitly and cannot be extracted directly from a model constraint.

Similarly, *negated cliques* [33] can be extracted from the problem. A negated clique is a set of binary variables of which at most one variable can be set to zero. However, for the ease of presentation, we transfer this back to the first case by introducing *negated variables* of the form $x_i' := 1 - x_i$. Negated variables are auxiliary variables and directly linked to the respective original variable, such that fixing a negated variable fixes the original one to the reverse value and vice versa. Thus, a negated clique is a clique on negated variables. Note that we also allow a mix of original and negated variables to be present in a clique. For the remainder of this article, we will not further discriminate between original and negated variables and assume cliques to be of the form given above.

In modern MIP solvers, the set of all detected cliques is stored in the so-called *clique table*. This global structure forms a relaxation of the MIP and is used by solver components, e.g., to create clique cuts [21] or to deduce stronger reductions in presolving and propagation [32]. In Sect. 4, we will show how the clique table can be used to define a neighborhood for a LNS based primal heuristic.

3.2 The Variable Bound Graph

Variable bound constraints are linear inequalities which contain exactly two variables. Depending on the sign of the coefficient, the variables bound each other. For example, a constraint $ax + by \geq c$ with $a > 0$ implies that x is bounded from

$$x - 2y \le 3 \quad (1)$$
$$x + 2z \le 2 \quad (2)$$
$$x + 3y \le 6 \quad (3)$$

(a) Constraint set.

$$x \le 2y + 3 \quad (1a)$$
$$y \ge \tfrac{1}{2}x - \tfrac{3}{2} \quad (1b)$$
$$x \le 2 - 2z \quad (2a)$$
$$z \le 1 - 0.5x \quad (2b)$$
$$x \le 6 - 3y \quad (3a)$$
$$y \le 2 - \tfrac{1}{3}x \quad (3b)$$

(b) Variable bounds.

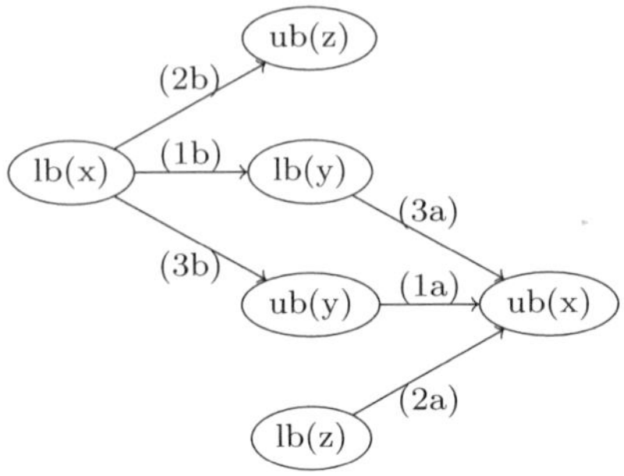

(c) Variable bound graph.

Fig. 1 Example of a variable bound graph

below by $\frac{c}{a} - \frac{b}{a}y$. If $a < 0$, the latter provides an upper bound on x. Consequently, a variable bound relation expresses the dependency of one bound of a variable on a bound of another variable. Typical examples for the use of variable bound constraints are precedence constraints on start time variables in scheduling or big-M constraints modeling fixed-costs in production planning.

Similar to the clique information, variable bound relations cannot only be deduced from variable bound constraints, but can also be identified within more general constraints or during presolving, e.g., by probing. Variable bound relations are exploited by different solver components, e.g., for c-MIR cut separation, where they can be used to replace non-binary variables with binary variables [27]. In order to make variable bound relations available for those components, they are stored in a global structure, the *variable bound graph*. In this directed graph, each node corresponds to the lower or upper bound of a variable and each variable bound relation is represented by an arc pointing from the influencing bound to the dependent bound. If a bound of a variable is tightened, implications can be read from this graph by following all paths starting at the corresponding node.

For an example of a variable bound graph, see Fig. 1. We regard three constraints on variable x, y, and z, as shown in part (a). Each of these constraints provides two bounds on the involved variables as stated in part (b). Thereby, bounds (1a) and (1b) are derived from constraint (1), (2a) and (2b) from (2), and (3a) and (3b) from (3). The resulting variable bound graph is illustrated in part (c). Each arc is labelled with the bound it represents.

4 Structure-Based Primal Heuristics

In this section, we present two new primal heuristics for mixed integer programming which are based on global structures collected by MIP solvers, see Sect. 3. Both heuristics are explicitly designed for use inside a MIP solver and not as standalone procedures. They employ the LNS paradigm (see Sect. 2) and are start heuristics, i.e., they do not need a feasible solution as reference point but aim at constructing an initial feasible solution.

In this aspect, they are similar to RENS; however, they differ in the definition of the neighborhood. While RENS uses a solution to the LP relaxation as reference point, our novel heuristics can even run before an LP is solved: They analyze the clique table and the variable bound graph built during the presolving phase and derive a set of variable fixings with a high probability of success for the subsequent LNS call.

The general scheme is the same for both heuristics. It is illustrated in Algorithm 1. In a first step, a subset of the integer variables is fixed based on the respective structure (lines 2–5). This is done in a diving-like fashion, i.e., in each iteration, the global structure is used to decide on one variable to be fixed. After that, two rounds of domain propagation are performed to avoid trivial infeasibilities and apply implied bound changes on other variables—those contained in the global structure, but also other variables in the problem. This process is iterated until all discrete variables in the used global structure were fixed. It can be interpreted as a dive in the tree with a domain propagation call after each fixing instead of solving a linear program, which is similar to the shift-and-propagate heuristic [10]. If domain propagation detects an infeasibility for the current assignment of variables, we backtrack one level, i.e., we undo the last fixing as well as the domain reductions deduced from it. Then, we remove the fixing value that led to the infeasibility from the domain of the respective variable and propagate this reduction. This is done in lines 6–14, and uses a method `domain_propagation`, which—given a MIP and the number of propagation rounds—performs domain propagation and returns the updated MIP as well as the information whether an infeasibility was detected during propagation. After the first backtrack, we stop the fixing process in order to avoid too much effort being spent by repeated backtracking.

If sufficiently many variables were fixed in this first phase, we solve an LP on the remaining problem and try to round the LP solution with a simple rounding heuristic [1] (see lines 16–19). Note that this LP is significantly smaller (and hopefully easier to solve) than the original LP relaxation because of the fixing threshold.

If the LP solution and the simple rounding heuristic did not lead to a primal feasible solution, the LNS search is started. For this, the neighborhood is defined by the fixings obtained in the previous phase. A sub-MIP for this neighborhood is created and solved (with a number of working limits to restrict the computational effort), see line 21. If a feasible solution was found during sub-MIP solving, it is returned.

Algorithm 1: Structure-based heuristics – general scheme

input : - MIP $\mathcal{P}(c, A, b, \ell, u, \mathcal{I})$
- fixing threshold α
- clique table $\mathcal{T}$ (NULL for variable bound heuristic), or
- variable bound graph $\mathcal{G}$ (NULL for clique heuristic)

output : - feasible solution or NULL, if no solution was found

```
begin
    // 1. fixing phase
    if T ≠ NULL then
        // apply Alg. 2 to fix variables
        (P(c, A, b, \tilde{ℓ}, \tilde{u}, I), inf) = clique_fixing(P(c, A, b, ℓ, u, I), T);
    else
        // apply Alg. 3 to fix variables
        (P(c, A, b, \tilde{ℓ}, \tilde{u}, I), inf) = variable_bound_fixing(P(c, A, b, ℓ, u, I), G);

    // 2. backtracking
    if inf then
        P(c, A, b, \tilde{ℓ}, \tilde{u}, I) ← undo last fixing x_k = f_k and induced propagations;
        // remove f_k from domain of x_k
        if \tilde{ℓ}_k = f_k then
            \tilde{ℓ}_k = \tilde{ℓ}_k + 1;
        else
            \tilde{u}_k = \tilde{u}_k − 1;
        // perform 2 rounds of domain propagation
        (P(c, A, b, \tilde{ℓ}, \tilde{u}, I), inf) = domain_propagation(P(c, A, b, \tilde{ℓ}, \tilde{u}, I), 2);
        if inf then
            return NULL;

    // 3. LP solving
    if |{i ∈ I | ℓ_i < u_i}| ≤ α|I| then
        x* ← solve P(c, A, b, \tilde{ℓ}, \tilde{u}, ∅);
        // try to round LP solution
        x* ← simple_round(x*);
        if x*_i ∈ ℤ for all i ∈ I then
            return x*;
        else
            // 4. sub-MIP solving
            x* ← solve P(c, A, b, \tilde{ℓ}, \tilde{u}, I) (with working limits, see Sect. 5);
            return x*;
    else
        return NULL;
end
```

The difference between the two heuristics is how and in which order the variables are fixed in the first step. The *clique heuristic* uses the clique table, while the *variable bound heuristic* takes into account the variable bound graph. Both these structures represent interconnections between variables which can and will be propagated. The novel concept of the heuristics is that the order in which variables are fixed and the fixing values take into account the predicted impact a fixing will have on the domain propagation step. By this, domain propagation is not used as a supplementary subroutine to support the search, but as a driving mechanism to take decisions within the search: we choose fixings of which we know that they propagate well. How this is done for the two named global structures is explained in the following.

4.1 Using the Clique Structure to Define Variable Fixings

In order to define a set of variable fixings with high probability of both being feasible and leading to a good solution, the clique structure can be used as illustrated in Algorithm 2.

First, a set of cliques is computed which partitions the set of binary variables. Note that a single binary variable always forms a trivial clique, so this partition is guaranteed to exist. Additionally, each clique cover can be transformed into a clique partition, since every subset of a clique forms a clique itself. Using a clique partition is a heuristic approach to finding a promising fixing order.

Given a clique of the partition, we first check if a variable within the clique was fixed to one already, e.g., by domain propagation in a previous iteration (lines 5–9). If this is not the case, we choose a variable with smallest objective coefficient among the unfixed variables in the clique and fix it to one, see lines 10–17. After that, two rounds of domain propagation are performed, which fixes all other unfixed variables in the clique to zero, but might also identify valid bound changes for variables in other cliques or even for variables not contained in the clique table. This is repeated until the propagation detects infeasibility or all cliques were handled. Since fixing a variable in a clique to one causes multiple other fixings, this scheme helps to reduce the neighborhood size. On the other hand, we set the cheapest variable in the clique to one in order to not increase the objective value too much and thereby aim at finding high quality solutions.

After all binary variables have been fixed by this algorithm, the remaining problem is solved by an LNS approach as illustrated in Algorithm 1.

Algorithm 2: `clique_fixing`

```
input  : - MIP P(c, A, b, ℓ̃, ũ, I)
         - clique table T
output : - subproblem P(c, A, b, ℓ̃, ũ, I) to solve as sub-MIP
         - Bool inf – TRUE if selected assignement was detected to
           be infeasible
begin
    // 1. clique partition
    compute clique partition C̄ = {C_1, ..., C_k};
    for i ∈ {1, ..., |C̄|} do
        fixed ← FALSE;
        // look for variable fixed to 1, e.g., by propagation of
           previous cliques
        for j ∈ C_i do
            if ℓ̃_j = 1 then
                if fixed then
                    // two variables fixed to 1 in a clique →
                       infeasible
                    return (P(c, A, b, ℓ̃, ũ, I), TRUE);
                fixed ← TRUE;
        if fixed = FALSE then
            // fix cheapest variable to 1
            x_{c_1}, ..., x_{c_{n_i}} ← sort variables of clique C_i by increasing objective coefficient;
            for j ∈ {1, ..., n_i} do
                if ũ_{c_j} = 0 then
                    continue ;
                else
                    ℓ̃_{c_j} ← 1;
                    break ;
        // perform 2 rounds of domain propagation
        (P(c, A, b, ℓ̃, ũ, I), inf) = domain_propagation(P(c, A, b, ℓ̃, ũ, I), 2);
        if inf then
            return (P(c, A, b, ℓ̃, ũ, I), TRUE);
    return (P(c, A, b, ℓ̃, ũ, I), FALSE);
end
```

4.2 *Using the Variable Bound Graph to Define Variable Fixings*

In the variable bound heuristic, we implemented different rules for determining the variable fixings. All of them make use of an (almost) topological sorting of the variable bound graph. A topological sorting of an acyclic directed graph is an order of the nodes, such that for every arc (i, j) node i precedes node j in the order. Since the variable bound graph can contain cycles, we may need to break them by randomly removing one of the arcs in the cycle. We call a topological sorting of this reduced graph *almost topological* and use this sorting to define the order in which variables are fixed. The fixing algorithms of the variables bound heuristics are summarized in Algorithm 3. As before, this is a sub-algorithm of Algorithm 1 and defines the subproblem to be solved by an LNS approach in a subsequent step.

The nodes of the variable bound graph are processed in the almost topological order, skipping continuous variables since we only want to fix integer variables (line 5). Additionally, variables fixed by domain propagation in a previous iteration are ignored, see line 6. Each node v of the variable bound graph represents a bound of a variable. Tightening this bound causes some bound changes on other variables, as defined by all paths in the variable bound graph starting at node v. Consequently, the earlier a variable bound is considered within the almost topological order, the more impact on other bounds we expect when tightening it.

All four variants process the nodes of the variable bound graph in almost topological order, but use different rules to decide whether and to which value the variable corresponding to a node is fixed (see lines 7–21).

The first variant by which the variable bound heuristic determines fixings aims at obtaining a large neighborhood by fixing variables such that only few additional restrictions are caused. This results in a neighborhood with a higher probability both for containing feasible solutions as well as high-quality solutions. To this end, this variant fixes the variable to the bound represented by the current node. This means that not the bound corresponding to the current node is tightened, but the opposite bound, which comes later in the topological order (if even) and thus causes fewer reductions on other bounds. In Algorithm 3, this variant is obtained when parameters `tighten` and `obj` are both set to `FALSE`.

The second variant uses an opposing argument: A large neighborhood is more expensive to process and finding any solutions in there might need more effort than in a smaller neighborhood with more fixed variables. Therefore, we fix the variable to the reverse bound, i.e., tighten the bound corresponding to the node in the variable bound graph. This forces change on many other variable bounds, a concept known to be rather effective in order to drive the solution to feasibility faster, cf. [29]. In Algorithm 3, this corresponds to `tighten` = `TRUE` and `obj` = `FALSE`.

We obtained two more variants by extending the previous ones to take into account the objective function (triggered by setting `obj` to `TRUE` in Algorithm 3). For this, we need the notion of the best bound of a variable, which is the bound that leads to the best objective contribution of the variable, i.e., its lower bound if its objective value is nonnegative, and its upper bound otherwise.

Variant three is based on variant one, in the sense that it fixes variables to the bound provided by the node in the topological sorting in order to obtain a large neighborhood. It even goes one step further: The variables are never fixed to their best bound, but only to their worst bound. While increasing the objective function value, these decisions are often less probable to lead to an infeasible subproblem, so that this variant aims mainly at finding a feasible solution.

Algorithm 3: `variable_bound_fixing`

input :
- MIP $\mathcal{P}(c, A, b, \tilde{\ell}, \tilde{u}, \mathcal{I})$
- variable bound graph $\mathcal{G}$
- Bool: `tighten` – should as much change as possible be caused?
- Bool: `obj` – should the objective function be taken into account?

output :
- subproblem $\mathcal{P}(c, A, b, \tilde{\ell}, \tilde{u}, \mathcal{I})$ to solve as sub-MIP
- Bool `inf` – `TRUE` if selected assignement was detected to be infeasible

begin
  `// 1. topological ordering`
  $b_1, \ldots, b_k \leftarrow$ bounds of variables in almost topological order;
  `// 2. process bound in topological order`
  for $i \in \{1, \ldots, k\}$ **do**
    $x_j \leftarrow$ variable represented by b_i;
    `// skip continuous variables`
    if $j \notin \mathcal{I}$ **then continue** ;
    `// variable is already fixed`
    if $\tilde{\ell}_j = \tilde{u}_j$ **then continue** ;
    `//` b_i `represents lower bound`
    if $b_i = lb(x_j)$ **then**
      if *not* *obj* ***or*** $c_j < 0$ **then**
        `// tighten the regarded bound`
        if *tighten* **then**
          $\tilde{\ell}_j \leftarrow \tilde{u}_j$;
        `// fix to the regarded bound`
        else
          $\tilde{u}_j \leftarrow \tilde{\ell}_j$;
    `//` b_i `represents upper bound`
    else
      if *not* *obj* ***or*** $c_j \geq 0$ **then**
        `// tighten the regarded bound`
        if *tighten* **then**
          $\tilde{u}_j \leftarrow \tilde{\ell}_j$;
        `// fix to the regarded bound`
        else
          $\tilde{\ell}_j \leftarrow \tilde{u}_j$;
    `// perform 2 rounds of domain propagation`
    $(\mathcal{P}(c, A, b, \tilde{\ell}, \tilde{u}, \mathcal{I}), \texttt{inf}) = \texttt{domain\_propagation}(\mathcal{P}(c, A, b, \tilde{\ell}, \tilde{u}, \mathcal{I}), 2)$;
    if *inf* **then**
      return $(\mathcal{P}(c, A, b, \tilde{\ell}, \tilde{u}, \mathcal{I}), \mathit{TRUE})$;
  return $(\mathcal{P}(c, A, b, \tilde{\ell}, \tilde{u}, \mathcal{I}), \mathit{FALSE})$;
end

On the other hand, the fourth variant tries to cause many changes by the fixings—similar to variant two—, so it changes the bound corresponding to the current node in the topological order by fixing the variable to the reverse bound. However, it only applies this fixing if this helps to improve the objective function value, i.e., if the variable is fixed to is its best bound. This allows for better solutions, while hopefully still fixing enough variables to have reasonable LNS solving times.

After each fixed variable, two rounds of domain propagation are performed (line 2) to identify consequences of the fixing.

There are arguments for each of these rules and indeed, none of them dominates the other but rather do they complement each other. Therefore, the variable bound heuristic is run up to four times in a row, once with each rule.

5 Computational Results

In this section, we present computational experiments showing the effect of the clique and variable bound heuristics. We used an implementation based on the academic MIP solver SCIP 3.1.1 [2] with SoPlex 2.0.1 [34] as underlying LP solver. All results were obtained on a cluster of 3.2 GHz Intel Xeon X5672 CPUs with 12 MB cache and 48 GB main memory. Each job was run exclusively on one node with a time limit of 3600 s.

Our experiments were performed on the MMMC test set which contains all instances from the last three MIPLIB benchmark sets [3, 12, 22] as well as the Cor@l test set [14]. We removed duplicates and the instances lrn, neos-1058477, neos-847051, and npmv07 because they caused numerical troubles with both variants. This left us with a total of 495 instances.

Each heuristic is implemented in a primal heuristics plugin of SCIP and called once at the beginning of the root node processing. Note that finding new incumbent solutions is often most effective at the root node, when a new primal bound might directly lead to global fixings, tighter cutting planes and better initial branching decisions. We decided not to run our global structure-based heuristics on local bounds, since the computational overhead of LNS heuristics is more significant than for, e.g., simple rounding or diving heuristics.

During the fixing phase of both heuristics, we apply all default domain propagation algorithms of SCIP, but limit the number of domain propagation rounds performed after a variable fixing by 2. This typically suffices to identify most implied bound changes and infeasibilities while avoiding spending too much time for propgation. In order to limit the effort spent within the heuristics, we use working limits for the subsequent LP and sub-MIP solving. First, we aim at having a significantly easier problem after the fixing phase. To this end, we run the heuristic only if at least 30 % of the integer variables were fixed. Second, we aim at performing a quick partial solve of the sub-MIP. Therefore, we disable separation in the LNS sub-MIP solving, use only fast presolving algorithms, and disable all LNS heuristics to avoid recursion. Additionally, we disable strong branching and use the inference branching rule of

Table 1 Root node statistics for the clique and variable bound heuristic individually and jointly on the MMMC test set (495 instances)

heuristic	affected models				unaffected models		
	instances	found	time (s)	share (%)	instances	time (s)	share (%)
clique	259	139	1.19	3.43	236	0.03	0.25
vbound	127	64	1.63	7.12	368	0.00	0.01
both	309	176	1.67	5.80	186	0.04	0.30

SCIP [1]. If a primal feasible solution was found already, we set an objective limit such that the solution is improved by at least 1 %. Finally, a node limit of 5000 is used together with a limit of 500 for the number of stalling nodes, i.e., consecutively processed nodes without finding a new best solution.

In our first experiment, we evaluated the effectiveness of the new heuristics and their running times. All other SCIP heuristics were disabled for this experiment. The results are presented in Table 1 for the two heuristics individually, as well as for both of them together. In the columns labeled "affected models", we list the number of instances on which the fixing rate of 30 % was reached, the number of instances on which a solution was found, and the average running time of the heuristic on these models. Note that the classification is done individually for all three variants and the sets of affected instances differ significantly so that the times should not be compared directly. Therefore, we also show the running time compared to the overall presolving and root node processing time in column "share (%)". Additionally, we also present the number of unaffected instances as well as the average running time and running time share on these instances in columns "unaffected models".

Our first observation is that the needed structures occur regularly: More than half of the instances contain a large enough clique structure to reach the desired fixing rate; on the other hand, the variable bound graph is sufficiently large for more than 25 % of the instances. Both heuristics succeed in finding a feasible solution on more than every second instance with the appropriate structure. Together, they find solutions for more than one third of the complete test set.

The running times on the affected instances are below two seconds on average for both heuristics individually as well as jointly. When compared to the root node processing time including presolving, the relative running time of the variable bound heuristic is twice as high as that of the clique heuristic which is consequential since it runs four different fixing rules (some of which might be fast because they do not lead to the desired fixing rate). On average over all affected instances, the running time of both heuristics together was less than 6 % of the overall root node solving time. This is more than reasonable for heuristics with a joint success rate of more than 50 %.

On the other hand, the running times on unaffected models are very small. The variable bound heuristic spends almost no time, while computing the clique partition in the clique heuristic can need up to half a second on some instances. On average,

Table 2 Solution process statistics for SCIP with default settings and with additional structural heuristics (clique and variable bound heuristic)

variant	MMMC (495 instances)					all optimal (299)	
	solved	time first	avg. gap	nodes	time (s)	nodes	time (s)
default	300	11.3	6.93 %	5892	340.4	1459	66.0
+ struct	305	10.6	6.58 %	5626	328.9	1347	63.0

however, the time needed is still quite small and both heuristics together increase the root node processing time by only 0.3 % on average for unaffected models.

In order to evaluate the impact of the newly proposed structure-based heuristics on the solving process, we performed another experiment which is summarized in Table 2. We compare the default settings of SCIP—with all default SCIP heuristics—to a variant with both structural heuristics enabled. Again, we observe an improvement by the structural heuristics. With clique and variable bound heuristic, we are able to solve 6 instances that were not solved before, while only one instance cannot be solved anymore. We reduce the time to finding the first solution (column "time first") by 6 % as well as the solving time by 3 % in the shifted geometric mean.[1] The average primal gap[2] (on instances for which a feasible solution is known) is decreased by 5 % as well, showing the consistent improvement in terms of solutions found and their quality. When looking only at those 299 instances solved to optimality by both variants, the time reduction amounts to 5 % and we see that also the number of branch-and-bound nodes needed to prove optimality is reduced by 8 % in the shifted geometric mean.

6 Conclusions and Outlook

In this paper, we presented two primal heuristics which are based on global structures available within MIP solvers, namely the clique table and the variable bound graph.

Based on these structures, we derive variable fixings which are applied iteratively with intermediate domain propagation rounds. The LP relaxation of the resulting subproblem is then solved. If the rounded LP solution is not feasible, the subproblem is solved in an LNS fashion. In our approach, domain propagation is not only used as a tool to avoid infeasible fixings, but rather are the fixing order and the fixing values decided based upon their effect on the domain propagation step. The global structures provide the tools to predict this effect by representing a part of the domain reductions that can be deduced from a variable fixing.

[1]For a definition and discussion of the shifted geometric mean, see [1, Appendix A]. We use shifts of 10 and 100 for time and nodes, respectively.

[2]We compute the average primal gap by means of the primal integral [7] as $P(t_{\max})/t_{\max}$ with $t_{\max} = 3600$ s, $P(x) = \int_{t=0}^{x} \gamma(t)\,dt$ and $\gamma(t)$ the primal gap at time t.

When applied carefully, these heuristics find solutions for more than one third of the instances in standard MIP benchmark sets. They slightly speed up the overall solving process and help to solve more instances to optimality. Therefore, clique and variable bound heuristic will be part of the next SCIP release.

Acknowledgments The work for this article has been conducted within the *Research Campus Modal* funded by the German Federal Ministry of Education and Research (fund number 05M14ZAM). The authors would like to thank the anonymous reviewer for helpful comments on the paper.

References

1. Achterberg, T.: Constraint integer programming. Ph.D. thesis, Technische Universität Berlin (2007)
2. Achterberg, T.: SCIP: solving constraint integer programs. Math. Program. Comput. **1**(1), 1–41 (2009)
3. Achterberg, T., Koch, T., Martin, A.: MIPLIB 2003. Oper. Res. Lett. **34**(4), 1–12 (2006). doi:10.1016/j.orl.2005.07.009
4. Achterberg, T., Raack, C.: The MCF-separator: detecting and exploiting multi-commodity flow structures in MIPs. Math. Program. Comput. **2**(2), 125–165 (2010)
5. Achterberg, T., Wunderling, R.: Mixed integer programming: analyzing 12 years of progress. In: Facets of Combinatorial Optimization, pp. 449–481. Springer (2013)
6. Berthold, T.: Primal heuristics for mixed integer programs. Diploma thesis, Technische Universität Berlin (2006)
7. Berthold, T.: Measuring the impact of primal heuristics. Oper. Res. Lett. **41**(6), 611–614 (2013)
8. Berthold, T.: Heuristic algorithms in global MINLP solvers. Ph.D. thesis, Technische Universität Berlin (2014)
9. Berthold, T.: RENS—the optimal rounding. Math. Program. Comput. **6**(1), 33–54 (2014)
10. Berthold, T., Hendel, G.: Shift-and-propagate. J. Heuristics **21**(1), 73–106 (2015)
11. Bixby, R.E.: A brief history of linear and mixed-integer programming computation. Documenta Mathematica pp. 107–121 (2012)
12. Bixby, R.E., Ceria, S., McZeal, C.M., Savelsbergh, M.W.P.: An updated mixed integer programming library: MIPLIB 3.0. Optima **58**, 12–15 (1998)
13. Borndörfer, R., Grötschel, M., Jäger, U.: Planning problems in public transit. In: Grötschel, M., Lucas, K., Mehrmann, V. (eds.) Production Factor Mathematics, pp. 95–121. Springer, Berlin (2010)
14. COR@L: MIP Instances (2014). http://coral.ie.lehigh.edu/data-sets/mixed-integer-instances/
15. Dakin, R.J.: A tree-search algorithm for mixed integer programming problems. Comput. J. **8**(3), 250–255 (1965)
16. Danna, E., Rothberg, E., Pape, C.L.: Exploring relaxation induced neighborhoods to improve MIP solutions. Math. Program. **102**(1), 71–90 (2004)
17. Fischetti, M., Lodi, A.: Local branching. Math. Program. **98**(1–3), 23–47 (2003)
18. Fischetti, M., Lodi, A.: Heuristics in mixed integer programming. In: J.J. Cochran, L.A. Cox, P. Keskinocak, J.P. Kharoufeh, J.C. Smith (eds.) Wiley Encyclopedia of Operations Research and Management Science. Wiley (2010)
19. Ghosh, S.: DINS, a MIP improvement heuristic. In: Fischetti, M., Williamson, D.P. (eds.) 12th International IPCO Conference, Proceedings of the Integer Programming and Combinatorial Optimization, LNCS, vol. 4513, pp. 310–323. Springer, Berlin (2007)
20. Heinz, S., Ku, W.Y., Beck, J.: Recent improvements using constraint integer programming for resource allocation and scheduling. In: Gomes, C., Sellmann, M. (eds.) Integration of AI and

OR Techniques in Constraint Programming for Combinatorial Optimization Problems. Lecture Notes in Computer Science, vol. 7874, pp. 12–27. Springer, Berlin (2013)
21. Johnson, E.L., Padberg, M.W.: Degree-two inequalities, clique facets, and biperfect graphs. North-Holland Math. Stud. **66**, 169–187 (1982)
22. Koch, T., Achterberg, T., Andersen, E., Bastert, O., Berthold, T., Bixby, R.E., Danna, E., Gamrath, G., Gleixner, A.M., Heinz, S., Lodi, A., Mittelmann, H., Ralphs, T., Salvagnin, D., Steffy, D.E., Wolter, K.: MIPLIB 2010. Math. Program. Comput. **3**(2), 103–163 (2011)
23. Land, A.H., Doig, A.G.: An automatic method of solving discrete programming problems. Econometrica **28**(3), 497–520 (1960)
24. Lee, E., Lewis, D.: Integer programming for telecommunications. In: Resende, M., Pardalos, P. (eds.) Handbook of Optimization in Telecommunications, pp. 67–102. Springer, US (2006)
25. Lodi, A.: Mixed integer programming computation. In: Jünger, M., Liebling, T.M., Naddef, D., Nemhauser, G.L., Pulleyblank, W.R., Reinelt, G., Rinaldi, G., Wolsey, L.A. (eds.) 50 Years of Integer Programming 1958–2008, pp. 619–645. Springer, Berlin (2010)
26. Lodi, A.: The heuristic (dark) side of MIP solvers. In: Talbi, E.G. (ed.) Hybrid Metaheuristics, Studies in Computational Intelligence, vol. 434, pp. 273–284. Springer, Berlin (2013)
27. Marchand, H., Wolsey, L.A.: Aggregation and mixed integer rounding to solve MIPs. Oper. Res. **49**(3), 363–371 (2001). doi:10.1287/opre.49.3.363.11211
28. Pochet, Y., Wolsey, L.A.: Production Planning by Mixed Integer Programming. Springer Science and Business Media, Heidelberg (2006)
29. Pryor, J., Chinneck, J.W.: Faster integer-feasibility in mixed-integer linear programs by branching to force change. Comput. Oper. Res. **38**(8), 1143–1152 (2011)
30. Rothberg, E.: An evolutionary algorithm for polishing mixed integer programming solutions. INFORMS J. Comput. **19**(4), 534–541 (2007)
31. Salvagnin, D.: Detecting and exploiting permutation structures in MIPs. In: Simonis, H. (ed.) Integration of AI and OR Techniques in Constraint Programming. Lecture Notes in Computer Science, vol. 8451, pp. 29–44. Springer, Berlin (2014)
32. Savelsbergh, M.W.P.: Preprocessing and probing techniques for mixed integer programming problems. ORSA J. Comput. **6**, 445–454 (1994)
33. Winkler, M.: Presolving for pseudo-Boolean optimization problems. Diploma thesis, Technische Universität Berlin (2014)
34. Wunderling, R.: Paralleler und objektorientierter Simplex-algorithmus. Ph.D. thesis, Technische Universität Berlin (1996)

Optimal Turbine Allocation for Offshore and Onshore Wind Farms

Martina Fischetti, Matteo Fischetti and Michele Monaci

Abstract Green energy became a topic of great interest in the recent years, as every year the demand for energy is increasing and the old resources, fossil fuels in particular, are pollutant and are becoming scarce. As a result, many countries have ambitious plans regarding green energy production. In particular, lots of money and energy are spent on the optimal design of wind farms, as an efficient use of the available resources is instrumental for their economical success. In the present paper we address the optimization of turbine positions, which is one of the most relevant problems in the design of a wind farm, and propose a heuristic approach based on Mixed-Integer Linear Programming techniques. Computational results on very large scale instances prove the practical viability of the approach.

Keywords Wind farm optimization · Heuristics · Mixed integer linear programming

1 Introduction

Energy optimization became a topic of great interest in the recent years, as the demand for energy is increasing every year. In addition, non-renewable resources (e.g., coal, natural gas or petroleum) are pollutant and are becoming scarce, hence a great attention has been posed on green energies. As a result, many countries such as Germany, the United Kingdom and Denmark invest lots of money and effort in research about green energy, and wind energy production in particular. Denmark, for example, has announced in February 2011 that 35 % of the energy produced will be renewable by 2020, and 100 % by 2050. As to the wind energy, the Danish wind

M. Fischetti
DTU Copenhagen, and Vattenfall BU Renewables, Denmark, Europe
e-mail: martifi@dtu.dk

M. Fischetti (✉) · M. Monaci
DEI, University of Padova, via Gradenigo 6/a, 35100 Padova, Italy
e-mail: matteo.fischetti@unipd.it

M. Monaci
e-mail: michele.monaci@unipd.it

K. Fujisawa et al. (eds.), *Optimization in the Real World*,
Mathematics for Industry 13, DOI 10.1007/978-4-431-55420-2_4

turbine industry is the world's largest and 90 % of the wind turbines manufactured in Denmark are sold to international markets.

An efficient use of the available resources is instrumental for the economical success of wind farms. In the present article we address the optimization of turbine positions, which is one of the most relevant problems in the design of a wind farm, and propose a heuristic approach based on Mixed-Integer Linear Programming techniques.

The Mathematical Programming approach for designing optimization tools for real-world problems consists in the definition of a sound mathematical model for the problem at hand, whose solution is initially attempted by means of some general-purpose solver used as a black box. However, mathematical models for wind farm optimization turn out to be very hard to solve to proven optimality, and even deriving high-quality heuristic solutions can be very problematic for the size and complexity of the models arising in practical cases. As a result, alternative solution techniques have to be derived for the sake of producing a viable tool that produces good heuristic solutions within acceptable computing times.

In our study we follow the approach known as *Matheuristic* [1] that hybridizes Mathematical Programming with Metaheuristics (hence its name). In particular, we develop different models and heuristics for the problem, and combine them within an overall proximity search [2] framework whose performance is computationally evaluated on real-world test cases.

The paper is organized as follows. The wind farm optimization problem is described in Sect. 2, and modeled through suitable Mixed-Integer Linear Programming (MIP) models. Simple ad-hoc heuristics that are not explicitly based on the MIP models are outlined in Sect. 3. The proximity search paradigm for general MIPs is outlined in Sect. 4, and used in Sect. 5 to derive a mathematically sound heuristic for wind-farm optimization. Computational results on offshore data are presented in Sect. 6, while the onshore case is outlined in Sect. 7. Possible extensions are finally addressed in Sect. 8.

The present article is intended to be as self-contained as possible. For more details the reader is referred to [2], where the proximity search approach is introduced and computationally evaluated on some relevant classes of MIPs, and to [3, 4] where the optimal turbine location problem is studied.

2 Wind Farm Layout Optimization

In wind systems, energy is generated by a set of turbines that convert the kinetic energy of the wind to the blades. All turbines are connected to the electric network that is used to collect all the energy. A main aspect to consider when addressing these systems in the actual position of turbines, as the impact of the wind against the turbines produces a decrease of the wind speed immediately behind the rotor. Thus, if another turbine is placed immediately downstream the first turbine, a reduction in the incoming wind speed—and hence a reduction in power production—is experienced.

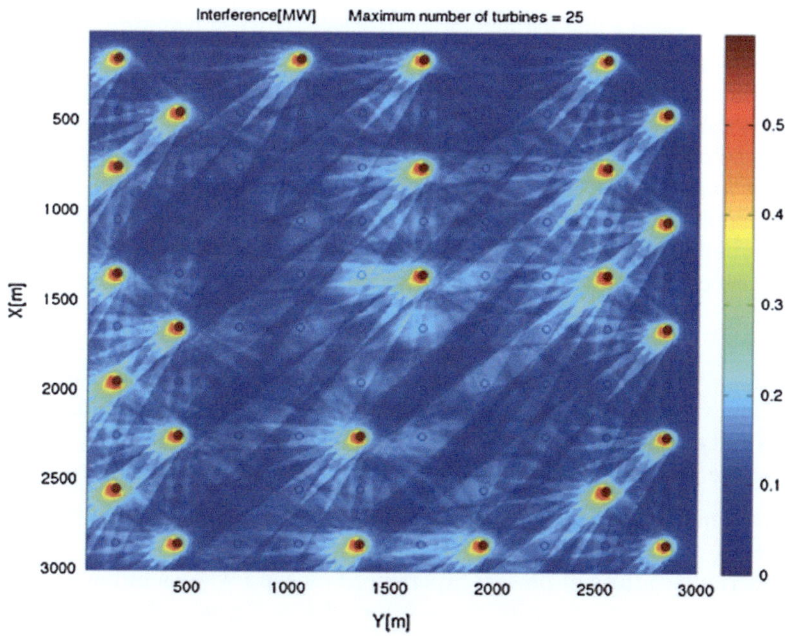

Fig. 1 Turbine packing in an offshore setting (100 potential sites in a 10×10 grid), with cumulative interference due to the combined action of different wind scenarios

In large offshore wind farms this phenomenon, known as "turbine wake", may reduce by 10–20 % the total energy production. It is then clear that an optimized layout of the turbines can have a very large impact in reducing the wake effect and, consequently, in terms of energy production.

Figure 1 illustrates a wind farm problem corresponding to a 3,000 $\times$ 3,000 (m) offshore area where turbines can be installed. The small circles identify the points where a turbine can potentially be built (*sites*), while filled circles refer to the actually built turbines. Interference due to the built turbines is represented in the background of the figure, and refers to the average of a large number of different wind scenarios (prevalent wind coming from North-East).

2.1 A Basic MIP Model

A basic MIP model from the literature is first reviewed, that focuses on turbine proximity constraints and on the wake effect; see, e.g., [5]. In particular, we consider the following constraints:

(a) a minimum and maximum number of turbines that can be built is imposed;
(b) a minimal separation distance is required between any pair of turbines; these constraints, known as *turbine distance constraints*, ensure that the blades do not physically clash.

Moreover, we assume that the mutual interference between two turbines will cause a loss in the power production that depends on their relative position and on wind conditions, and that the total interference can be approximated as the sum of pairwise interferences.

We denote by $V = \{1, \dots, n\}$ the set of possible positions for a turbine, called "sites" in what follows, and by D_{MIN} the minimum distance between two turbines. After computing the (symmetric) distance $dist(i, j)$ between each pair of sites i and j, one can define the incompatibility graph $G_I = (V, E_I)$ that has one vertex for each site and one edge for each pair of sites that are mutually incompatible, namely

$$E_I = \{[i, j] : i, j \in V,\ dist(i, j) < D_{MIN},\ i < j\}.$$

In addition, we denote by N_{MIN} and N_{MAX} the minimum and maximum number of turbines that can be built, respectively.

Given the technological specifications of the turbines, one can also compute

- the maximum power, denoted as P_i, that a turbine would produce if built (alone) at site i;
- the interference (loss of power) experienced by site j when a turbine is installed at site i, denoted by I_{ij} (with $I_{jj} = 0$ for all $j \in V$); note that matrix I is not symmetrical due to wind direction.

To model the problem, we introduce two sets of binary variables:

$$x_i = \begin{cases} 1 & \text{if a turbine is built at site } i; \\ 0 & \text{otherwise} \end{cases} \qquad (i \in V) \tag{1}$$

$$z_{ij} = \begin{cases} 1 & \text{if two turbines are built at both sites } i \text{ and } j; \\ 0 & \text{otherwise} \end{cases} \qquad (i, j \in V,\ i < j) \tag{2}$$

The model then reads:

$$\max \sum_{i \in V} P_i x_i - \sum_{i \in V} \sum_{j \in V : i < j} (I_{ij} + I_{ji})\, z_{ij} \tag{3}$$

$$\text{s.t.} \quad N_{MIN} \le \sum_{i \in V} x_i \le N_{MAX} \tag{4}$$

$$x_i + x_j \le 1 \qquad \forall [i, j] \in E_I \tag{5}$$

$$x_i + x_j - 1 \le z_{ij} \qquad \forall i, j \in V,\ i < j \tag{6}$$

$$x_i \in \{0, 1\} \qquad \forall i \in V \tag{7}$$

$$z_{ij} \in \{0, 1\} \qquad \forall i, j \in V,\ i < j \tag{8}$$

The net power production, to be maximized in the objective function (3), takes the total power production and the interference among sites into account. Constraint (4) imposes a lower and upper bound on the number of turbines to be allocated, while inequalities (5) model the turbine distance constraints. Finally, constraints (6) force

$z_{ij} = 1$ whenever $x_i = x_j = 1$, whereas the objective function ensures that $z_{ij} = 0$ whenever $x_i = 0$ and/or $x_j = 0$, i.e., in any optimal solution we have $z_{ij} = x_i x_j$, as required.

Note that constraints (5) could be strengthened to their clique counterpart

$$\sum_{h \in Q} x_h \le 1 \qquad \forall Q \in \mathcal{Q} \tag{9}$$

where $\mathcal{Q}$ is a family of maximal cliques of G_I, such that every edge in E_I is contained in at least one member of $\mathcal{Q}$. In our implementation, we decided not to improve constraints (5) to their clique form (9), for two main reasons: (i) the number of cliques can be exceedingly large, and (ii) a family of cliques is automatically generated during preprocessing by most of the state-of-the-art MIP solvers, which can control their number and can handle them very efficiently. As we have no additional information about which cliques should be generated explicitly, we preferred to leave clique management to the solver itself.

A relevant feature of the above model is that it allows for an easy extension to an uncertain setting where the definition of the turbine power vector P and of interference matrix I depends on the wind scenario considered—as wind direction and speed greatly vary in time, this is very important for the practical accuracy of the model. Using statistical data, one can indeed collect a large number K of wind scenarios k, each associated with a pair (P^k, I^k) and with a probability π_k, and write the following straightforward Stochastic Programming variant of the previous model: the objective function becomes

$$z = \sum_{k=1}^{K} \pi_k \left(\sum_{i \in V} P_i^k x_i - \sum_{i \in V} \sum_{j \in V : i < j} (I_{ij}^k + I_{ji}^k) z_{ij} \right) \tag{10}$$

while all constraints stay unchanged as they only involve "first-stage" variables x and z. It is therefore sufficient to define

$$P_i := \sum_{k=1}^{K} \pi_k P_i^k \qquad \forall i \in V \tag{11}$$

$$I_{ij} := \sum_{k=1}^{K} \pi_k I_{ij}^k \qquad \forall i, j \in V \tag{12}$$

to obtain the same model (3)–(8) as before.

Though alternative proposals are available in the literature (see, e.g., [6]), we decided to stick to the basic model above for two main reasons: (i) the model is quite standard and well understood by practitioners, (ii) as already mentioned, a suitable definition of the input data allows one to easily take different wind scenarios into account, whereas more sophisticated models would lead to really huge Stochastic Programming variants.

2.2 A Handier MIP Model

While (3)–(8) turns out to be a reasonable model when just a few sites have to be considered (say $n \approx 100$), it becomes hopeless to solve when $n \geq 1{,}000$ because of the huge number of variables and constraints involved, that grows quadratically with n. Therefore, when facing instances with several thousand sites, one has to resort to an alternative (possibly weaker) model where interference can be handled by a number of variables and constraints that grows just linearly with n. Following a recipe of Glover [7] which is widely used, e.g., in the Quadratic Assignment Problem [8], one can rewrite the objective function as

$$\sum_{i \in V} P_i x_i - \sum_{i \in V} \Big(\sum_{j \in V} I_{ij} x_j\Big) x_i \tag{13}$$

and restate it as

$$\sum_{i \in V} (P_i x_i - w_i) \tag{14}$$

where

$$w_i := \Big(\sum_{j \in V} I_{ij} x_j\Big) x_i = \begin{cases} \sum_{j \in V} I_{ij} x_j & \text{if } x_i = 1; \\ 0 & \text{if } x_i = 0. \end{cases} \qquad (i \in V) \tag{15}$$

is a new variable that denotes the total interference caused by site i. Then, a compact model reads:

$$\max \sum_{i \in V} (P_i x_i - w_i) \tag{16}$$

$$\text{s.t.} \quad N_{MIN} \leq \sum_{i \in V} x_i \leq N_{MAX} \tag{17}$$

$$x_i + x_j \leq 1 \qquad \forall [i,j] \in E_I \tag{18}$$

$$\sum_{j \in V} I_{ij} x_j \leq w_i + M_i (1 - x_i) \qquad \forall i \in V \tag{19}$$

$$x_i \in \{0, 1\} \qquad \forall i \in V \tag{20}$$

$$w_i \geq 0 \qquad \forall i \in V \tag{21}$$

where the big-M term $M_i = \sum_{j \in V : [i,j] \notin E_I} I_{ij}$ is used to deactivate constraint (19) in case $x_i = 0$.

3 Simple Ad-Hoc Heuristics

A simple 1-opt heuristic can be designed along the following lines. Assume that, at each step, we have an incumbent solution, say $\tilde{x}$, that describes the best-known turbine allocation ($\tilde{x}_i = 1$ if a turbine is built at site i, 0 otherwise), and a current solution x. Let

$$z = \sum_{i \in V} P_i x_i - \sum_{i \in V} \sum_{j \in V} I_{ij} x_i x_j$$

be the profit of the current solution, $\gamma = \sum_{i \in V} x_i$ be its cardinality, and define for each $j \in V$ the extra-profit δ_j incurred when flipping x_j, namely:

$$\delta_j = \begin{cases} P_j - \sum\limits_{i \in V: x_i = 1} (I_{ij} + I_{ji}) & \text{if } x_j = 0; \\ -P_j + \sum\limits_{i \in V: x_i = 1} (I_{ij} + I_{ji}) & \text{if } x_j = 1 \end{cases}$$

where we assume $I_{ij} = BIG$ for all incompatible pairs $[i, j] \in E_I$, and BIG is a large penalty value (e.g., $BIG > \sum_{i \in V} P_i$), while $I_{ii} = 0$ as usual.

We start with $x = 0$, $z = 0$, $\gamma = 0$ and initialize $\delta_j = P_j$ for all $j \in V$. Then, we iteratively improve x by a sequence of 1-opt moves, each changing the value of one single variable, according to the following scheme. At each iteration, we look in $O(n)$ time for the site j with maximum $\delta_j + FLIP(j)$, where function $FLIP(j)$ takes cardinality constraints into account, namely

$$FLIP(j) = \begin{cases} -HUGE & \text{if } x_j = 0 \text{ and } \gamma \geq N_{MAX} \\ -HUGE & \text{if } x_j = 1 \text{ and } \gamma \leq N_{MIN} \\ +HUGE & \text{if } x_j = 0 \text{ and } \gamma < N_{MIN} \\ +HUGE & \text{if } x_j = 1 \text{ and } \gamma > N_{MAX} \\ 0 & \text{otherwise} \end{cases}$$

with $HUGE \gg BIG$ (recall that the figure $\delta_j + FLIP_j$ has to be maximized).

Once the best j has been found, say j^*, if $\delta_{j^*} + FLIP(j^*) > 0$ we just flip x_{j^*}, update x, z, and γ in $O(1)$ time, update all δ_j's in $O(n)$ time (through the parametric technique described in [4]), and repeat. In this way a sequence of improving solutions is obtained, until a local optimal solution that cannot be improved by just one flip is found. (To escape local minima, a simple perturbation scheme can be implemented; see again [4] for details.)

We also implemented a 2-opt heuristic to allow a single turbine to move to a better site—a move that requires flipping two variables. Each 2-opt exchange requires $O(n^2)$ time as it amounts to trying n 1-opt exchanges and to apply the best one.

The above ad-hoc heuristics are applied in two different modes. In the "initial solution" mode, we start with $\tilde{x} = x = 0$ and repeat the procedure (including perturbation to escape local optimal solutions) until we count a very large number

(10,000) of consecutive 1-opt calls with no improvement of $\tilde{x}$. In the faster "clean-up" mode, instead, we already have a hopefully good incumbent $\tilde{x}$ to refine, so we initialize $x = \tilde{x}$ and repeat the procedure until we count 100 consecutive 1-opt calls with no improvement of $\tilde{x}$. As to the time-consuming 2-opt exchanges, they are applied with a certain frequency, and in any case just before the final $\tilde{x}$ is returned.

4 Proximity Search Heuristics for General Mixed-Integer Programs

Proximity search is a general approach aimed at improving a given feasible "reference solution", quickly producing a sequence of solutions of improved quality. For the sake of being self-contained, we next outline the main features of this technique, and refer the reader to [2] for fuller details.

Proximity search is related to *Large-Neighborhood Search* (LNS) heuristics [9], that also explore a solution neighborhood defined by invalid constraints. For instance, *Local Branching* [10] adds a constraint that limits the search to solutions that are "sufficiently close" to the reference solution. Similarly, *Relaxation Induced Neighborhood Search* (RINS) [11] is a heuristic that addresses the neighborhood resulting from fixing all variables having the same value in the reference and in relaxation solutions. Proximity search is also related to the parametric branch-and-bound algorithm proposed by Glover [12] and extended in [13] to a parametric tabu search algorithm. A different approach that defines a neighborhood of a given solution is the *Feasibility Pump* paradigm introduced in [14] for 0-1 Mixed-Integer Linear Programs, and extended to nonlinear problems in [15, 16] (among others).

A distinguished feature of proximity search is that no invalid constraint is added to the model, but the objective function is modified to favor solutions that are "close enough" to the reference solution. The idea proved quite effective in quickly improving a given starting feasible solution, at least when the landscape of feasible solutions is not too irregular—as it happens in the wind farm layout optimization context.

4.1 The Basic Idea

For the sake of generality, in this section we will focus on a generic 0-1 Mixed-Integer (possibly nonlinear) Program of the form

$$\min f(x) \tag{22}$$

$$g(x) \le 0 \tag{23}$$

$$x_j \in \{0, 1\} \;\; \forall j \in J \tag{24}$$

Algorithm 1 The basic Proximity Search heuristic

1: let $\tilde{x}$ be the initial heuristic feasible solution to refine;
2: **while** an overall termination condition is not reached **do**
3: explicitly add the *cutoff constraint* $f(x) \leq f(\tilde{x}) - \theta$ to the MIP model;
4: replace $f(x)$ by the proximity objective function $\Delta(x, \tilde{x})$;
5: run the MIP solver on the new model until a termination condition is reached, and let x^* be the best feasible solution found (x^* empty if none);
6: **if** x^* is nonempty and $J \subset N$ **then**
7: refine x^* by solving the convex program $x^* := \text{argmin}\{f(x) : g(x) \leq 0,\ x_j = x_j^*\ \forall j \in J\}$
8: **end if**
9: recenter $\Delta(x, \cdot)$ by setting $\tilde{x} := x^*$, and/or update θ
10: **end while**
11: **return** $\tilde{x}$

where $f : \Re^n \to \Re$, $g : \Re^n \to \Re^m$, and $J \subseteq N := \{1, \ldots, n\}$, $J \neq \emptyset$, indexes binary variables. Although this is not strictly required by the method, in the following we assume that both f and g are convex functions with the property that dropping the integrality condition in (24) leads to a polynomially solvable relaxation.

Proximity search starts with a feasible solution $\tilde{x}$, and modifies the MIP formulation as follows:

- add an explicit *cutoff constraint*

$$f(x) \leq f(\tilde{x}) - \theta, \tag{25}$$

  where $\theta > 0$ is a given cutoff tolerance; and
- replace the original objective function with the Hamming distance

$$\Delta(x, \tilde{x}) := \sum_{j \in J : \tilde{x}_j = 0} x_j + \sum_{j \in J : \tilde{x}_j = 1} (1 - x_j) \tag{26}$$

The conceptual scheme of proximity search is sketched in Algorithm 1.

Proximity search requires an initial solution $\tilde{x}$ at Step 1. This solution can be computed by using some ad-hoc heuristic or by running the black-box MIP solver from scratch until a first feasible solution is found. In any case, we assume that finding a feasible solution is not really an issue for the problem at hand. If this is not the case, one should resort to a problem reformulation where some constraints are imposed in a soft way through violation penalties attached to slack variables.

At Step 3, the cutoff tolerance θ is defined and a the cutoff constraint is added.

At Step 4, the objective function is redefined as the Hamming distance between x and $\tilde{x}$, see (26). Then, the resulting problem is solved by using a black-box MIP solver to hopefully find a new feasible solution, say x^*. If this is the case, it must be $f(x^*) \leq f(\tilde{x}) - \theta$, i.e., x^* improves over $\tilde{x}$.

A key property of the approach is that the root-node solution of the convex relaxation, say x', is expected to be not too different from $\tilde{x}$, as this latter solution would be optimal without the cutoff constraint. As a matter of fact, for a small θ this constraint can typically be fulfilled with just minor adjustments of $\tilde{x}$, a property that is instrumental for the success of the method because of two main positive effects:

- the computing time spent at the root node is often very small (even orders of magnitude smaller than the time required at the root node for the original problem);
- solution x' is typically "almost integer", i.e., with a small number of fractional components indexed by J, thus improving the chances of success of the MIP internal heuristics.

Table 1 is taken from [2], and illustrates both positive effects for the set covering (pure binary) MIPLIB2010 instance `ramos3` when considering a reference solution $\tilde{x}$ of value 267. The table reports, for different values of parameter θ, the number of components of the LP relaxation solution x' that belong to the intervals [0,0], (0, 0.1], …(0.9, 1], and [1,1], along with computing time (in CPU sec.s), number of simplex iterations (dual pivots), and objective value—i.e., distance $\Delta(x', \tilde{x})$. The LP relaxation becomes infeasible for $\theta > 121$. The table shows that, for small values of θ, the LP-solution time is just negligible, while the number of integer components is very large. On the contrary, using a too large value for θ leads to LP-solutions that are "far away" from $\tilde{x}$ in terms of distance, have a large number of fractional components, and require a considerable computational effort to be computed—thus vanishing all proximity search benefits.

If no new solution x^* is found at Step 5 (possibly because the MIP solver was aborted before convergence), one proceeds directly to Step 9 where tolerance θ is reduced. Of course, if the MIP solver proved infeasibility for the given θ, one has that $f(\tilde{x}) - \theta$ is a valid lower bound on the optimal value of the original MIP.

Step 7 is aimed at improving the new solution x^*, if any, by solving a convex problem where all binary variables have been fixed to their value in x^*; in this way, the best solution within the neighborhood induced by $\Delta(x, x^*) = 0$ is determined.

At Step 9, the approach is iterated using the current incumbent $\tilde{x}$ (if available) so as to recenter the distance function Δ, and/or by modifying the cutoff tolerance θ.

4.2 Proximity Search Implementations

In this section we sketch three possible implementations of the basic proximity search method, as described in [2]. All three implementations start with a given solution $\tilde{x}$, replace the original objective function $f(x)$ by a proximity one $\Delta(x, \cdot)$, and use a cutoff constraint to force the detection of improved solutions.

In the first implementation, denoted as "without recentering" and described in Sect. 4.2.1, the proximity objective function $\Delta(x, \cdot)$ remains centered on the very first solution $\tilde{x}$, while the cutoff constraint is modified on the fly inside the MIP solver.

Table 1 Distribution of fractionalities in the LP relaxation solution x' and corresponding computing time and distance $\Delta(x', \tilde{x})$ from the reference solution $\tilde{x}$, for various values of the cutoff parameter θ

x-range	$\theta = 0$	$\theta = 1$	$\theta = 2$	$\theta = 3$	$\theta = 4$	$\theta = 5$	$\theta = 10$	$\theta = 20$	$\theta = 30$	$\theta = 50$	$\theta = 99$	$\theta = 121$
= 0	1920	1919	1919	1919	1924	1920	1619	1619	1600	1565	1276	682
(0.0, 0.1]	0	0	0	0	0	0	303	297	293	281	420	926
(0.1, 0.2]	0	0	0	0	0	4	0	6	26	65	194	380
(0.2, 0.3]	0	1	0	5	0	0	0	3	7	15	64	169
(0.3, 0.4]	0	0	0	0	0	0	0	1	2	8	75	29
(0.4, 0.5]	0	0	6	0	0	0	8	4	3	16	91	0
(0.5, 0.6]	0	0	0	0	0	0	5	5	9	19	47	1
(0.6, 0.7]	0	0	0	0	0	0	0	2	9	35	17	0
(0.7, 0.8]	0	5	0	1	0	1	0	10	25	88	3	0
(0.8, 0.9]	0	0	0	0	0	11	0	28	101	68	0	0
(0.9, 1.0)	0	0	0	0	0	0	249	209	110	26	0	0
=1	267	262	262	262	263	251	3	3	2	1	0	0
time (sec.s)	0.00	0.04	0.03	0.03	0.04	0.21	0.45	0.54	0.57	0.90	4.77	30.91
# LP-iter.s	0	352	341	357	358	1180	2164	2543	2637	3627	6829	11508
Δ-distance	0.00	1.50	3.00	4.50	6.00	7.88	17.45	37.13	56.86	96.90	208.71	292.67

The second implementation, called "with recentering" (see Sect. 4.2.2), is an iterative scheme in which the MIP solver is halted as soon as an improved solution, say x', is found. In this case, a new problem is defined by replacing $\tilde{x}$ by x' both in the objective $\Delta(x, \cdot)$ and in the cutoff constraint, and the MIP solver is re-applied from scratch.

A variant of the second implementation is given in Sect. 4.2.3. We call it "with incumbent", as the cutoff constraint is imposed in a soft way to make the initial solution $\tilde{x}$ feasible—though highly penalized in the objective function. In this way the current incumbent $\tilde{x}$ can be used for a warm-start of the MIP solver.

4.2.1 Proximity Search Without Recentering

This version of proximity search assumes that the MIP solver can be controlled through a callback function to be executed each time the incumbent is going to be updated—this is a standard feature of modern MIP solvers. Within such a function, the new incumbent $\hat{x}$ (say) is first internally recorded in a user's data structure, and a new global cut

$$f(x) \leq f(\hat{x}) - \theta \tag{27}$$

is added to the current model. The new cutoff (27) makes solution $\hat{x}$ infeasible, thus preventing the solver to update its own incumbent, and forces search to explore only solutions improving over $\hat{x}$.

Inequality (27) is imposed for the initial solution $\tilde{x}$ as well, making it infeasible. In this way the optimal relaxation solution x' at the root node is different from $\tilde{x}$, and violated MIP cuts can possibly be generated at the root node.

The simple implementation above has a main advantage in that a single enumeration tree is generated. However, the are some main drawbacks that can affect the performance of the method in a negative way, namely:

- The proximity objective function is always computed with respect to the initial solution $\tilde{x}$ and is not changed during the search. Thus, $\Delta(x, \cdot)$ remains "centered" with the very first solution (hence the name "proximity search without recentering"), though after some enumeration one has to explore a solution space that is far away from $\tilde{x}$.
- The MIP incumbent is never explicitly updated, which prevents the application of powerful propagation and variable-fixing schemes—as well as refinement heuristics—embedded in the MIP solver.
- The scheme requires to install some callback functions, using the MIP solver as a grey-box, which may deactivate some of its features.

4.2.2 Proximity Search with Recentering

As already noticed, the scheme presented in the previous section has some drawbacks, mainly related to the need of interacting with the underlying MIP and with the choice to keep the objective function centered with the first solution $\tilde{x}$. We now present a different implementation that uses the MIP solver as a black box (with no need of callback functions) and just restarts it as soon as a new solution is found.

In the new implementation, called "proximity search with recentering", Steps 1–4 are the same as in Algorithm 1. After the new problem has been defined, one invokes the MIP solver as a black box (Step 5), in its default mode and without any callback, and aborts its execution as soon as a first feasible solution is found. Due to the cutoff constraint in the model, this solution (if any) is a strict improvement over $\tilde{x}$. At Step 9 the method is then iterated, by replacing $\tilde{x}$ with the new solution and repeating (without changing θ) from Step 3, until the overall time limit is reached. Obviously, in case no solution is found at Step 5, the algorithm either proves θ-optimality of the incumbent $\tilde{x}$ or hits the given time limit.

We observe that this scheme has the main advantage to use the MIP solver as a black-box (in its default settings, without callbacks), and that it implements a dynamic updating of the objective function. In addition, the scheme can be implemented very easily. A main disadvantage of the method is that overlapping search trees are explored, possibly wasting computing time and solving the (computationally heavy) root node of the problem many times.

4.2.3 Proximity Search with Incumbent

The third implementation we address is aimed at solving a drawback that is common for both implementations above; namely, adding the cutoff constraint prevents the MIP solver to update its internal incumbent and to apply some powerful refinement heuristics such as RINS [11]. In the "proximity search with recentering" variant one imposes the cutoff constraint (25) in its "soft version"

$$f(x) \leq f(\tilde{x}) - \theta + z \tag{28}$$

where $z \geq 0$ is a continuous slack variable. Variable z is highly penalized in the objective function, which is modified to

$$\Delta(x, \tilde{x}) + Mz \tag{29}$$

where M is a large positive value compared to the feasible values of Δ. With this formulation, solution $\tilde{x}$ can be provided on input to the MIP solver as a feasible warm-start solution with $z = 1$. Though the cost of this solution is very large, having an internal incumbent allows the MIP solver to trigger its internal refinement heuristics. In this case too, whenever a new incumbent is found with $z = 0$, i.e., a θ-improving solution has been found, the execution is aborted and the method is iterated.

5 A Proximity Search Heuristic for Wind Farm Optimization

In this section we describe a heuristic algorithm for wind farm layout optimization that exploits a MIP model to systematically improve a starting feasible solution. As our final goal is to attack real cases with a very large number of sites, our approach is built around the weak model (16)–(21), and uses on the proximity search methodology described in the previous section.

Our overall heuristic can be casted into the so-called *MIP-and-refine* framework recently investigated in [17], and works as in Algorithm 2. The approach makes use of very fast (1- and 2-opt) heuristics described in Sect. 3. It also requires a parameter $\overline{N}$ used to heuristically reduce the number of candidate sites (in our tests, $\overline{N}$ =2,000 was used).

Algorithm 2 The overall MIP-and-refine framework

1: read input data and compute the overall interference matrix (I_{ij});
2: apply ad-hoc heuristics (1- and 2-opt) to get a first incumbent $\tilde{x}$;
3: **while** time limit permits **do**
4: apply quick ad-hoc refinement heuristics (few iterations of 1- and 2-opt) to possibly improve $\tilde{x}$;
5: if $n > \overline{N}$, randomly remove points $i \in V$ with $\tilde{x}_i = 0$ so as to reduce the number of candidate sites to $\overline{N}$;
6: build a MIP model for the resulting subproblem and apply proximity search to refine $\tilde{x}$ until the very first improved solution is found (or time limit is reached);
7: **end while**
8: **return** $\tilde{x}$

At Step 2 (respectively, Step 4) the ad-hoc heuristic of Sect. 3 is applied in its initial-solution (resp., clean-up) mode.

The proximity search heuristic used at Step 6 follows the "proximity search with incumbent" implementation of Sect. 4.2.3, and operates on two different MIP models. In a preliminary phase, a simplified MIP model is obtained from (16)–(21) by removing all interference constraints (19), thus obtaining a much easier relaxation. During this phase, proximity search is executed with a short time limit (60 s, in our tests) so as to compute solutions in which the number of turbines built increases at each iteration, without being bothered by interference considerations and only taking pairwise incompatibility (18) into account. This scheme is able to quickly produce a sequence of better and better solutions (even in terms of the true profit), until either (i) no additional turbine can be built, or (ii) the addition of new turbines does in fact reduce the true profit associated to the new solution. In both cases we switch to the complete model (16)–(21) that includes interference constraints, and use it in all subsequent executions of Step 6. Observe that the simplified model is only used at Step 6, while all other steps of the procedure always use the complete model that takes interference into full account.

6 Computational Experiments on Offshore Cases

In this section we describe our computational experiments for the turbine layout problem in the offshore settings. In particular, we considered an offshore 3,000 × 3,000 (m) square region and imposed a minimum distance $D_{MIN} = 400$ (m) between each pair of turbines, with no explicit bounds on the number of turbines to be built (i.e., $N_{MIN} = 0$ and $N_{MAX} = +\infty$). Turbines are all of Siemens SWT-2.3-93 type (diameter 93 m), producing a power of 0.0 MW for wind speed up to 3 m/s, of 2.3 MW for wind speed greater than or equal to 16 m/s, and intermediate values for winds in range 3–16 m/s according to a nonlinear function [18]. Pairwise interference (in MW) was computed using Jensen's model [19], by averaging 250,000+ real-world wind samplings grouped into about 500 macro-scenarios. A pairwise average interference of 0.01 MW or less is treated as zero. The reader is referred to [3] for details.

We generated five classes of medium-to-large problems with number of sites n ranging from 1,000 to 20,000. For each value of n, we considered 10 instances by randomly generating n points in the 3,000 × 3,000 square according to a uniform distribution.

The following alternative solution approaches were implemented in C language, some of which used the commercial package IBM ILOG Cplex 12.5.1 [20] as a MIP solver. Because of the big-M's involved in the models, all Cplex's codes use a strict zero without numerical tolerances (`CPX_PARAM_EPINT` = 0.0).

(a) `proxy`: the MIP-and-refine heuristic, as outlined in the previous section, using Cplex with the following aggressive parameter tuning: all cuts deactivated, `CPX_PARAM_RINSHEUR` = 1, `CPX_PARAM_POLISHAFTERTIME` = 0.0, `CPX_PARAM_INTSOLLIM` = 2;
(b) `cpx_def`: the application of IBM ILOG Cplex 12.5.1 in its default setting, starting from the same heuristic solution $\tilde{x}$ found by `proxy` after the first execution of Step 2 of Algorithm 2;
(c) `cpx_heu`: same as `cpx_def`, with the following internal tuning intended to improve Cplex's heuristic performance: all cuts deactivated, `CPX_PARAM_RINSHEUR` = 100, `CPX_PARAM_POLISHAFTERTIME` = 20 % of the total time limit;
(d) `loc_sea`: a benchmark local-search procedure (written in C) not based on any MIP solver: we just loop on Step 2 of Algorithm 2 and randomly remove installed turbines from the current best solution after 10,000 iterations without improvement of the incumbent.

In our view, `cpx_def` and `cpx_heu` represent a standard way of exploiting a MIP model once a good feasible solution is known, whereas `loc_sea` is somehow representative of a clever but not over-sophisticated metaheuristic, as typically implemented by practitioners.

All algorithms were run on an Intel Xeon E3-1220 V2 quad-core PC with 16GB of RAM. The outcome of our computational experiments is reported in Table 2, where each entry refers to the performance of a given algorithm at a given time limit. In particular, the first (leftmost) part of the table reports, for each algorithm and time

Table 2 Number of times each algorithm finds the best solution within the time limit (wins), and optimality ratio with respect to the best known solution—the larger the better (from [4])

n	Time limit (s)	Number of wins				Optimality ratio			
		proxy	cpx_def	cpx_heu	loc_sea	proxy	cpx_def	cpx_heu	loc_sea
1,000	60	**6**	1	3	0	**0.994**	0.983	0.987	0.916
	300	**4**	2	**4**	0	0.997	0.991	**0.998**	0.922
	600	**7**	3	**7**	0	**0.997**	0.992	**0.997**	0.932
	900	**5**	2	3	0	**0.998**	0.993	0.996	0.935
	1,200	**5**	1	**5**	0	**0.998**	0.992	0.997	0.939
	1,800	**5**	1	4	0	**0.998**	0.992	0.996	0.942
	3,600	4	2	**5**	0	**0.998**	0.995	0.997	0.943
5,000	60	**9**	6	6	5	**0.909**	0.901	0.901	0.904
	300	**10**	0	0	0	**0.992**	0.908	0.908	0.925
	600	**10**	0	**10**	0	**0.994**	0.908	**0.994**	0.935
	900	**10**	0	0	0	**0.994**	0.908	0.908	0.936
	1,200	**10**	0	0	0	**0.994**	0.908	0.925	0.939
	1,800	**9**	0	1	0	0.996	0.908	**0.971**	0.946
	3,600	**5**	0	**5**	0	**0.996**	0.932	0.994	0.948
10,000	60	9	9	8	**10**	**0.914**	0.913	**0.914**	**0.914**
	300	**10**	2	2	2	**0.967**	0.927	0.927	0.936
	600	**10**	0	**10**	0	**0.998**	0.928	**0.998**	0.944
	900	**10**	0	0	0	**1.000**	0.928	0.928	0.948
	1,200	**10**	0	0	0	**1.000**	0.928	0.928	0.951
	1,800	**10**	0	0	0	**1.000**	0.928	0.928	0.957
	3,600	**9**	0	0	1	**1.000**	0.928	0.928	0.964

(continued)

Table 2 (continued)

n	Time limit (s)	Number of wins				Optimality ratio			
		`proxy`	`cpx_def`	`cpx_heu`	`loc_sea`	`proxy`	`cpx_def`	`cpx_heu`	`loc_sea`
15,000	60	9	**10**	9	9	0.909	**0.912**	0.911	0.909
	300	**10**	8	7	8	**0.943**	0.937	0.935	0.937
	600	**10**	0	**10**	0	**0.992**	0.939	**0.992**	0.942
	900	**10**	0	0	0	**1.000**	0.939	0.939	0.956
	1,200	**9**	0	0	1	**1.000**	0.939	0.939	0.959
	1,800	**9**	0	0	1	**1.000**	0.939	0.939	0.965
	3,600	**9**	0	0	1	**1.000**	0.939	0.939	0.972
20,000	60	9	9	9	**10**	0.901	**0.902**	0.901	**0.902**
	300	**10**	8	**10**	**10**	**0.933**	**0.933**	**0.933**	**0.933**
	600	**9**	0	**9**	1	**0.956**	0.935	**0.956**	0.941
	900	**10**	0	0	0	**0.978**	0.935	0.935	0.945
	1,200	**10**	0	0	0	**0.991**	0.935	0.935	0.950
	1,800	**10**	0	0	0	**0.999**	0.935	0.935	0.963
	3,600	**9**	0	0	0	**1.000**	0.935	0.935	0.971
All	60	**42**	35	35	34	**0.925**	0.922	0.922	0.909
	300	**44**	20	23	20	**0.966**	0.939	0.940	0.930
	600	**46**	3	**46**	1	**0.987**	0.941	**0.987**	0.938
	900	**45**	2	3	0	**0.994**	0.941	0.941	0.944
	1,200	**44**	1	5	1	**0.997**	0.940	0.945	0.947
	1,800	**43**	1	5	1	**0.999**	0.940	0.954	0.955
	3,600	**36**	2	10	2	**0.999**	0.946	0.959	0.959

limit, the *number of wins*, i.e., the number of instances for which the algorithm produced the best solution at the given time limit (ties allowed). All winning values in Table 2 are marked in bold.

According to the table, `proxy` outperforms all competitors by a large amount for medium to large instances. As expected, `cpx_heu` performs better for instances with $n = 1{,}000$ as it is allowed to explore a large number of enumeration nodes for the original model and objective function. Note that `loc_sea` has a good performance for short time limits and/or for large instances, thus confirming its effectiveness, whereas `cpx_heu` is significantly better than `loc_sea` only for small instances and large time limits.

The right-hand side part of Table 2 provides a different performance measure, and reports the *average optimality ratio*, i.e., the average value of the ratio between the solution produced by an algorithm (on a given instance at a given time limit) and the best solution known for that instance—the larger the better, as we have a maximization problem. It should be observed that an improvement of just 1 % has a very significant economical impact in the wind farm context. These results confirm that `proxy` is always able to produce solutions that are quite close to the best one. As before, `loc_sea` is competitive for large instances when a very small computing time is allowed, whereas `cpx_def` and `cpx_heu` exhibit a good performance only for small instances, and are dominated even by `loc_sea` for larger ones.

Figure 2 plots the incumbent value (i.e., the profit of the current best solution) over CPU time for the four heuristics under comparison, and refers to 4 sample instances

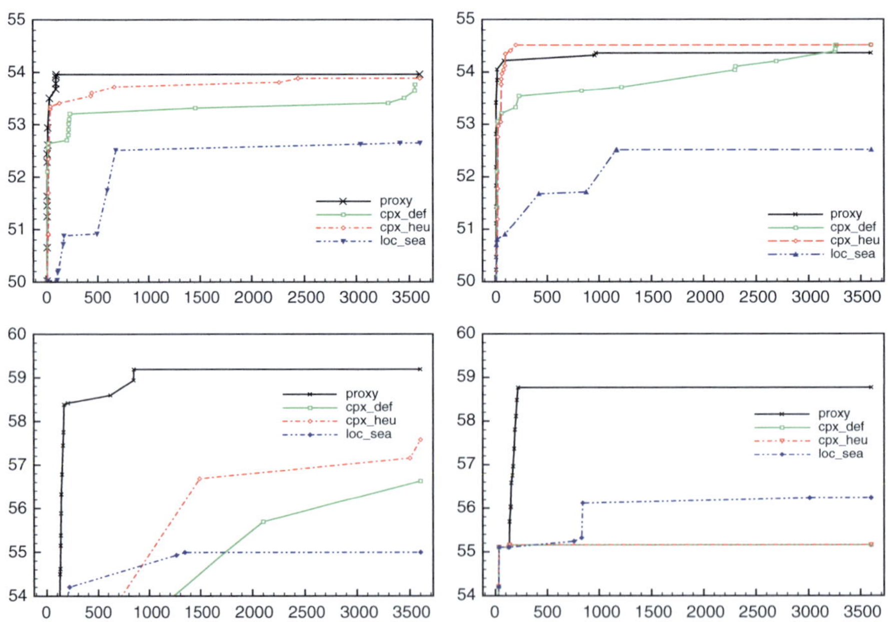

Fig. 2 Solution profit over time for 4 sample instances with $n = 1{,}000$ (*top left* and *top right*), $n = 5{,}000$ (*bottom left*), and $n = 10{,}000$ (*bottom right*); the higher the profit the better

in our testbed. The two subfigures on the top refer to two small instances with $n =$ 1,000, where `proxy`, `cpx_heu` and `cpx_def` have a comparable performance and clearly outperform `loc_sea`. For $n = 5{,}000$ (bottom-left subfigure) and $n =$ 10,000 (bottom-right subfigure), however, both `cpx_def` and `cpx_heu` (and also `loc_sea`) have hard time in improving their initial solution, and are outperformed by `proxy` by a large amount.

7 The Onshore Case

Onshore wind farm is a particularly interesting case for turbine optimization that refers to the construction of wind farms on lands. Due to the irregularities of the terrain (mountains, hills, valleys) new constraints and non-homogeneous wind conditions must be considered in the optimization.

Within this section we will refer, as an example, to a real-world case located in the United Kingdom (near Edinburgh: latitude 55.798624, longitude –2.431291), and depicted in Fig. 3.

The onshore case has two main peculiarities with respect to the offshore case:

1. different heights and presence of obstacles make the wind non uniform, even without any interference between turbines. Figure 4 plots average wind speed obtained by Computational Fluid Dynamics (CFD) simulations, showing that wind speed varies significantly from point to point.

Fig. 3 A 3D plot of the real-world landscape to be used as onshore test case

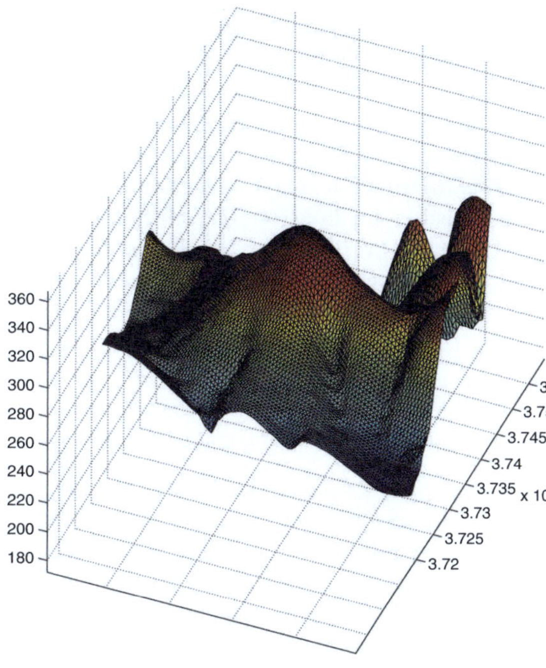

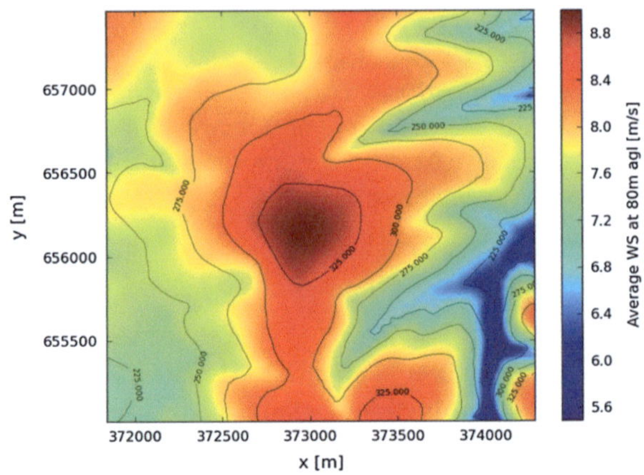

Fig. 4 Average wind speed [m/s] at 80 m agl (above ground level) for our test site

2. Not all the positions in the site are actually feasible due to terrain constraints including:

   - *Extreme wind*: If the wind is too strong, turbines risk to be damaged. Points where the average wind is too strong must be avoided.
   - *Turbulence*: Wind turbulence increases the load on the blades, gearbox, generator and tower. In addition, turbulent wind can excite resonances (large vibrations) in the turbine and in its supporting tower. All those loads can cause fatigue problems in turbine structures.
   - *Flow inclination*: When wind turbines are placed on steep slopes or cliffs, the wind might hit the rotor not perpendicularly. A large in-flow angle could lead to a high level of fatigue for the turbine.
   - *Shear coefficient*: The shear coefficient represents the presence of obstructions in the area, such as buildings, trees, etc. If the shear coefficient is too large a position cannot be used.
   - *Wind distribution compliance*: Zones where the wind is too strong could not be feasible positions for the selected turbines, according to the IEC 61400 standards. IEC 61400 is a set of wind-turbine standards required to ensure the engineering integrity of wind turbines, providing an appropriate level of protection against damage from all hazards during the planned lifetime. According to IEC, turbines are divided into classes depending on how much wind speed they can well sustain.
   - *Terrain inclination*: Excessive terrain inclination could be problematic for the installation of the turbine and can cause high fatigue on the turbine structure.

As a result, only a subset of positions is feasible for turbine allocation, as represented in dark blue in the combined constraint map reported in Fig. 5.

The combination of the two aforementioned effects is that a turbine can give different power productions depending on its position (even without considering any interference). With respect to the possible positions obtained by the previous analysis, the actual power distribution (without interference) is shown in Fig. 6.

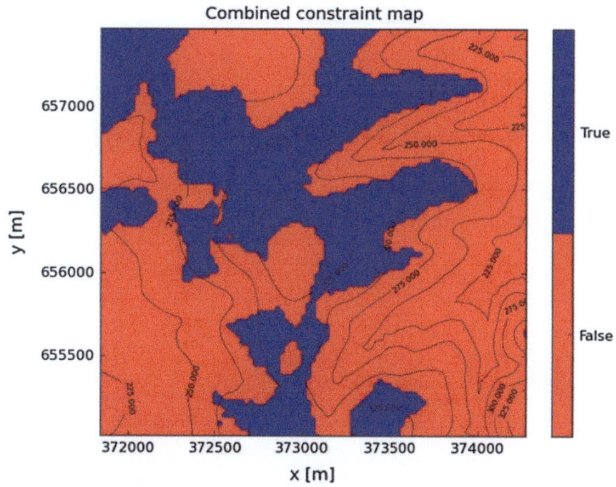

Fig. 5 Feasibility of turbine positions within our onshore site

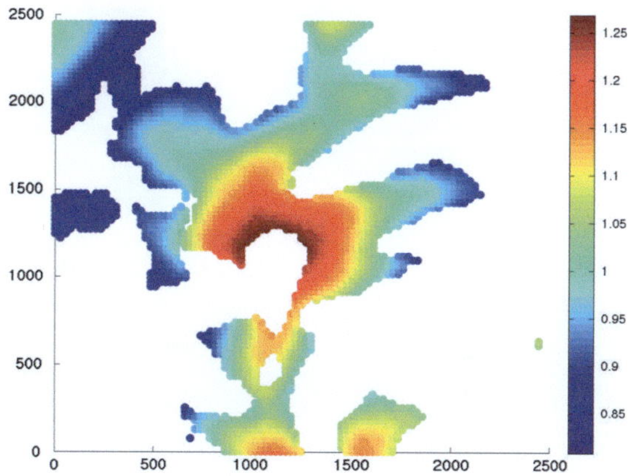

Fig. 6 Power distribution within our onshore area (feasible positions only)

The above considerations suggest that the onshore problem is no longer a 2D problem, but becomes a 3D one. The classical Jensen's model [19] for computing pairwise interferences can be adapted to a 3D case as suggested in [3]. In any case, MIP formulation (16)–(21) remains valid and can still be applied within the heuristic scheme described in Sect. 5.

In our onshore optimization, different wind scenarios have been obtained by considering wind directions only—wind intensity being fixed, for each feasible position, at the average wind speed reported in Fig. 4. This solution has been adopted because of the huge dimension of the problem: if one would consider all the different wind speed scenarios for each possible positions (250000+ winds for 3000+ positions) the size of the problem would grown dramatically. In addition, we assumed that wind direction does not change from point to point, though the actual wind tends to follow the shape of the ground.

Figure 7 reports the optimal allocation of turbines for different values of N_{MAX}; as expected, turbines tend to concentrate on the positions with more power production (cf. Fig. 6).

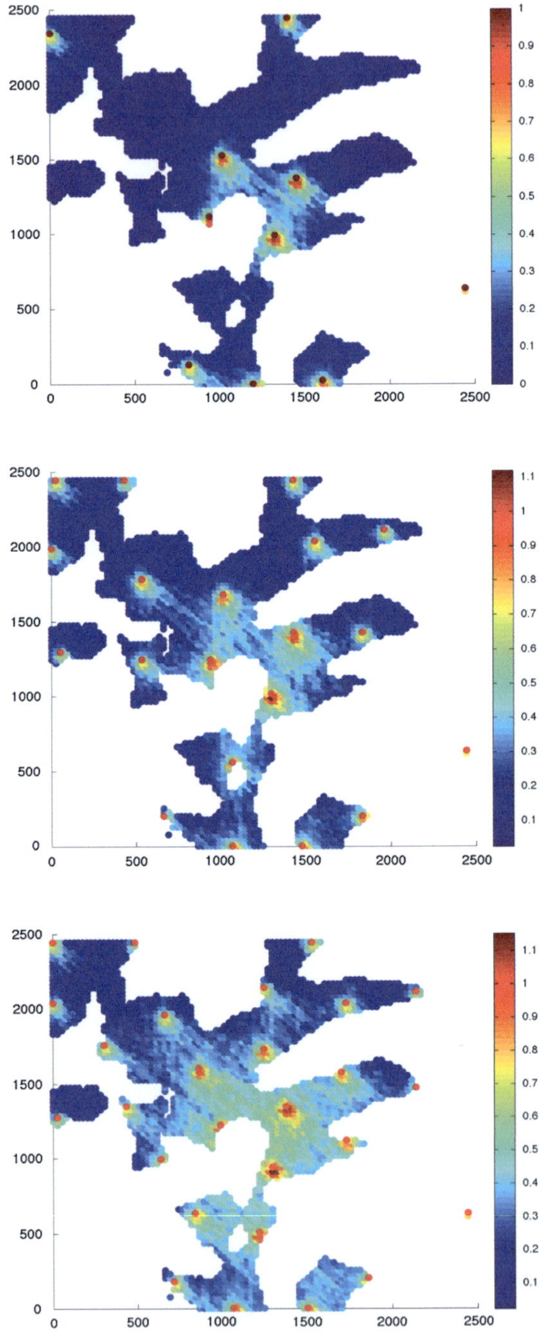

Fig. 7 Optimal turbine placements when considering a maximum number of turbines equal to 10, 20, and unlimited, resp.; *colors* in the background give the interference induced by the installed turbines (*red dots*)

8 Extensions

In this final section we give a sketch of possible extensions to the basic turbine location problem, as addressed in the previous sections. For all extensions, future research should be devoted to the evaluation the practical effectiveness of sound matheuristics such as proximity search on real word data.

Additional costs can be considered in the basic model, for the sake of reducing building costs, amount of required work and material, distance from roads and transformer stations, external power grid, maintenance costs, and so on. Additional constraints can similarly be added, including the load on the turbines or terrain conditions.

Also of interest is layout optimization when different types of turbines can be built. As a matter of fact, different turbine types exist, that differ because of the power they can handle, their dimensions and costs. As a general rule, the bigger the turbine the better, as big turbines can produce more power and the cost to build them is not much larger than the one for building smaller turbines. On the other hand, especially in onshore case, big turbines can not always be built in some areas, due to neighbor complains and natural restrictions. In addition, there are regulations imposing maximum sound levels originating from wind turbines at nearby houses or other sensitive areas, the sound level depending not only on the distance between the wind farm and the sensitive areas, but also on the number of built turbines and their dimension. It is therefore of practical interest to determine the optimal layout in a given area in the presence of different types of turbines and of the associated constraints.

A more challenging extension considers the electrical connection between the built turbines. In the offshore case, for example, all the power produced by the turbines is first collected in an offshore substation, and then transferred to the coast through cables. In [21] the resulting inter-array cable routing optimization is solved through a MIP-based approach. A natural next step is therefore to put together wind farm layout optimization and cable routing, addressing the two problems by a single model. Indeed, cable costs tend to create a compact layout, while interference tends to spread turbines all over the available area. A unified optimization approach would therefore be able to balance these two conflicting goals so as to obtain an efficient layout from both wind-resource and cable-cost points of view. Also of interest is to find the optimal position of the collecting substation as well, that should be at the same time as close as possible to the coast and to the turbines, and easily reachable by boats.

Acknowledgments The research of the first author was partially supported by Vattenfall BU Renewables and by the Denmark Innovation Fund. The remaining authors were supported by the University of Padova (Progetto di Ateneo "Exploiting randomness in Mixed Integer Linear Programming"), and by MiUR, Italy (PRIN project "Mixed-Integer Nonlinear Optimization: Approaches and Applications"). We thank Jesper Runge Kristoersen and Iulian Vranceanu from Vattenfall BU Renewables for their help in defining the cable routing constraints.

References

1. Maniezzo, V., Stützle, T., Voß, S. (eds.): Matheuristics—Hybridizing Metaheuristics and Mathematical Programming. Annals of Information Systems, vol. 10. Springer, New York (2010)
2. Fischetti, M., Monaci, M.: Proximity search for 0–1 mixed-integer convex programming. J. Heuristics **6**, 709–731 (2014)
3. Fischetti, M.: Mixed-integer models and algorithms for wind farm layout optimization. Master's thesis, University of Padova and Aalborg (2014)
4. Fischetti, M., Monaci, M.: Proximity search heuristics for wind farm optimal layout. Technical Report, to appear in Journal of Heuristics (2015)
5. Donovan, S.: Wind farm optimization. In: Proceedings of the 40th Annual ORSNZ Conference, pp. 196–205 (2005)
6. Archer, R., Nates, G., Donovan, S., Waterer, H.: Wind turbine interference in a wind farm layout optimization mixed integer linear programming model. Wind Eng. **35**(2), 165–178 (2011)
7. Glover, F.: Improved linear integer programming formulations of nonlinear integer problems. Manage. Sci. **22**, 455–460 (1975)
8. Fischetti, M., Monaci, M., Salvagnin, D.: Three ideas for the quadratic assignment problem. Oper. Res. **60**(4), 954–964 (2012)
9. Shaw, P.: Using constraint programming and local search methods to solve vehicle routing problems. In: Maher, M., Puget, J.-F. (eds.) Principles and Practice of Constraint Programming CP98. Lecture Notes in Computer Science, vol. 1520, pp. 417–431. Springer, Berlin (1998)
10. Fischetti, M., Lodi, A.: Local branching. Math. Program. **98**, 23–47 (2003)
11. Danna, E., Rothberg, E., Le Pape, C.: Exploring relaxation induced neighborhoods to improve MIP solutions. Math. Program. **102**(1), 71–90 (2005)
12. Glover, F.: Parametric branch and bound. OMEGA **6**(2), 145–152 (1978)
13. Glover, F.: Parametric tabu search for mixed integer programs. Comput. Oper. Res. **33**(9), 2449–2494 (2006)
14. Fischetti, M., Glover, F., Lodi, A.: The feasibility pump. Math. Program. **104**(1), 91–104 (2005)
15. Bonami, P., Cornuéjols, G., Lodi, A., Margot, F.: A feasibility pump for mixed integer nonlinear programs. Math. Program. **119**(2), 331–352 (2009)
16. D'Ambrosio, C., Frangioni, A., Liberti, L., Lodi, A.: A storm of feasibility pumps for nonconvex MINLP. Math. Program. **136**(2), 375–402 (2012)
17. Fischetti, M., Sartor, G., Zanette, A.: A MIP-and-refine matheuristic for smart grid energy management. Int. Trans. Oper. Res. 1–11 (2013)
18. Siemens AG: SWT-2.3-93 Turbine, Technical Specifications. http://www.energy.siemens.com
19. Jensen, N.: A note on wind generator interaction. Technical Report Riso-M-2411(EN), Riso National Laboratory, Roskilde, Denmark (1983)
20. IBM ILOG CPLEX: Optimization Studio. http://www.cplex.com
21. Fischetti, M., Leth, J.: A Mixed-Integer Linear Programming approach to wind farm layout and inter-array cable routing. Technical Report, University of Aalborg (2014)

Optimal Cycles for Persistent Homology Via Linear Programming

Emerson G. Escolar and Yasuaki Hiraoka

Abstract In this work, we discuss the problem of finding optimal cycles for homology groups of simplicial complexes and for persistent homology of filtrations. We review the linear programming formulation of the optimal homologous cycle problem and its extension to allow for multiple cycles. By inserting these linear programming problems into the persistent homology algorithm, we are able to compute an optimal cycle, that has been optimized at birth, for every persistent interval in the persistent diagram.

Keywords Computational homology · Persistent homology · Optimal cycles · Linear programming

1 Introduction

One of the techniques in computational topology for data analysis is the use of invariants from algebraic topology in order to study the topological features of data. In particular, the homology groups can be interpreted as containing information about topological features such as connected components, loops, voids, and so on. Defined algebraically, these homology groups are quotient groups, and thus the homology classes are determined by representative cycles. However, given a homology class, its representative cycle is not uniquely defined. From the point of view of applications, we are interested in finding "optimal" representatives. Moreover, this optimality should reflect the underlying geometry of the input.

E.G. Escolar (✉)
Graduate School of Mathematics, Kyushu University,
744 Motooka, Nishi-ku, Fukuoka 819-0395, Japan
e-mail: eescolar@math.kyushu-u.ac.jp

Y. Hiraoka
WPI—Advanced Institute for Materials Research (WPI-AIMR),
Tohoku University, 2-1-1, Katahira, Aoba-ku, Sendai 980-8577, Japan
e-mail: hiraoka@wpi-aimr.tohoku.ac.jp

K. Fujisawa et al. (eds.), *Optimization in the Real World*,
Mathematics for Industry 13, DOI 10.1007/978-4-431-55420-2_5

Given an input space, which we here take to be a simplicial complex, we compute its homology groups. Intuitively speaking, for example, given a homology class in dimension $q = 1$ indicating the presence of a hole (a loop) in the space, a representative can be a cycle that winds around the hole loosely. We would like to tighten the cycle around the hole so that it accurately captures not just the existence, but also the location of the hole.

Our basic approach follows [5], where they consider the optimal homologous cycle problem over $\mathbb{Z}$ coefficients using integer linear optimization. There, the focus is on a single homology class $[z]$ and computing an optimal representative for this class. With some assumptions on the input, they show that in this formulation, an optimal solution can be obtained in polynomial time via linear relaxation.

We do not restrict our attention to a single homology class, but rather consider homology bases and generators. A *homology basis* is a set of cycles $G = \{z_1, \ldots, z_n\}$ such that $\{[z_1], \ldots, [z_n]\}$ is a basis for the homology group (if it happens to be free). Analogously, a *homology generator* is a set of cycles G such that their homology classes generate the homology group.

We use an iterative procedure that optimizes each cycle representative z_i, factoring out the other cycles in G. This produces a new homology generator (or basis) G' where each representative has been optimized. Moreover, by using this iterative process, we can easily extend our formulation to persistent homology.

Recently, the use of persistent homology [8, 17] has become popular for topological data analysis. Persistent homology captures how topological features persist through an increasing sequence of spaces. The persistence diagram summarizes the topological information encoded in the birth-death intervals. To each feature we associate its birth-death interval, representing its lifespan in the sequence. In this view, interesting patterns may appear in the persistence diagram itself. For example, see the study [15].

To further elucidate the features encoded in the persistence diagram, it is important to be able to identify, for each persistence interval, a corresponding representative cycle. Moreover, these cycles should be chosen to be optimal. As in the single simplicial complex case, we use linear optimization to define and compute this optimality. Here, persistent homology requires field coefficients so we drop the integrality conditions.

The optimization algorithm can be inserted directly into the persistence algorithm. One way to compute persistence is to process the simplices in birth order, iteratively. Adding to this, in each iteration we update our optimization problem corresponding to the entry of a new variable representing this simplex. Moreover, whenever a cycle is born, we optimize it immediately, performing optimization at the birth of a feature.

The problems of computing optimal cycles and optimal bases in homology groups have been extensively studied. Even though the basic goal, which is to get optimal representatives, is the same, there are many different possible formulations due to the choice of underlying field and varying definitions of optimality. Below is a short overview, but we do not attempt to give a complete list.

In the case of dimension $q = 1$, the problem of computing the shortest homology basis over field coefficients has a polynomial time algorithm [9] using a greedy algorithm. This is further improved in one of the results of [6] which uses persistence to prune candidates in the greedy algorithm. These methods produce a shortest homology basis, which is defined in the following way. Each cycle z is given a weight $w(z)$ equal to its number of simplices, and the shortest homology basis problem is to find a homology cycle basis G that minimizes $w(G) = \sum w(z_i)$. For dimension $q > 1$, the paper [3] shows that over $\mathbb{Z}_2$ coefficients, the shortest homology basis problem is NP-hard.

The paper [4] consider a slightly different problem, defining the weight of a homology class $h = [z]$ as the size of the smallest ball that contains it. They define a minimum homology basis as a basis that minimizes total weight with respect to this definition, and provide a polynomial time algorithm for computing such a basis.

Over $\mathbb{Z}$ coefficients, we have already mentioned the paper [5] which considers the optimal homologous cycle problem using linear programming. We chose this formulation for its simplicity and flexibility.

For ease of visualization, all of our examples will be given in dimension $q = 1$. We emphasize, however, that the discussion applies for any dimension $q \geq 1$.

2 Background

Here, we review some basic concepts. We refer the reader to [14], for example, for algebraic topology. For integer programming, we refer to the book [16], for example.

2.1 *Simplicial Homology*

Roughly speaking, a simplicial complex is a set of vertices, edges, faces, and so on, together with information about how these parts are glued together. Formally, we define a simplex to be a finite set $S = \{v_0, \ldots, v_q\}$ of vertices. Any subset of S is called a face of S. A *simplicial complex* X is simply a collection of simplices, such that the face of any simplex in X is also a simplex in X. In this work we consider only simplicial complexes with a finite number of simplices.

An orientation of a simplex $S = \{v_0, \ldots, v_q\}$ is an equivalence class of orderings of the vertices of S, where two orderings are equivalent if they differ by an even permutation. That is, the two orderings $v_0, \ldots, v_q$ and $v_{\pi(0)}, \ldots, v_{\pi(q)}$ are equivalent if and only if the permutation π has sign $+1$. To every simplex in X, we assign an orientation and denote an oriented simplex by $[v_0, \ldots, v_q]$. The dimension of an oriented simplex $\sigma = [v_0, \ldots, v_q]$ is q. The set of all q-dimensional oriented simplices of X is denoted by X_q.

Let X be a simplicial complex. We define its qth chain group as the group generated by the q-dimensional oriented simplices. In other words, we have

$$C_q(X) = \{\sum_{\sigma \in X_q} n_\sigma \sigma | n_\sigma \in \mathbb{Z}\}.$$

Elements of $C_q(X)$ are called q-chains of X. Let π be a permutation on $\{0, \ldots, q\}$. We equate $[v_0, \ldots, v_q] = \mathrm{sgn}(\pi)[v_{\pi(0)}, \ldots, v_{\pi(q)}]$.

In this setting, we use coefficients in $\mathbb{Z}$. More generally, it is possible to use coefficients in some other commutative ring R by replacing $\mathbb{Z}$ with R in this definition.

We also have the boundary maps $\partial_q : C_q(X) \to C_{q-1}(X)$ defined via linear extension of

$$\partial_q \sigma = \sum_{i=0}^{q} (-1)^i [v_0, \ldots, \hat{v_i}, \ldots, v_q]$$

defined on q-simplices $\sigma = [v_0, \ldots, v_q]$, where the $\hat{v_i}$ means that the vertex v_i is omitted. For example, in the simplicial complex in Fig. 1, we have $\sigma = [v_0, v_1, v_2]$, and so $\partial_2 \sigma = [v_1, v_2] - [v_0, v_2] + [v_0, v_1] = e_2 + e_0 + e_1$.

Let $Z_q(X) = \ker \partial_q$ and $B_q(X) = \mathrm{im}\ \partial_{q+1}$. The chains in $Z_q(X)$ are called q-cycles, while those in $B_q(X)$ are called q-boundaries. The qth homology group of X is defined to be the quotient group

$$H_q(X) = \frac{Z_q(X)}{B_q(X)} = \frac{\ker \partial_q}{\mathrm{im}\ \partial_{q+1}}.$$

Given $z \in Z_q(X)$, we denote its homology class in $H_q(X)$ by $[z] = z + B_q(X)$. Moreover, two cycles $z, z' \in Z_q(X)$ are said to be homologous, denoted by $z \sim z'$, if and only if they are in the same homology class; that is, if and only if $[z] = [z']$.

In general, we talk about chains, cycles, boundaries and homology groups without reference to the dimension q, as it should be clear from context. We warn that in this context, the term cycles is not limited to the usual meaning of "cycles" in graph theory.

By choosing a basis on each $C_q(X)$, $q \geq 0$, each ∂_q can be written in matrix form. The computation of $H_q(X)$ can be performed by matrix reduction, specifically by computing the Smith normal forms of the ∂_q. We refer the reader to [13] for more details.

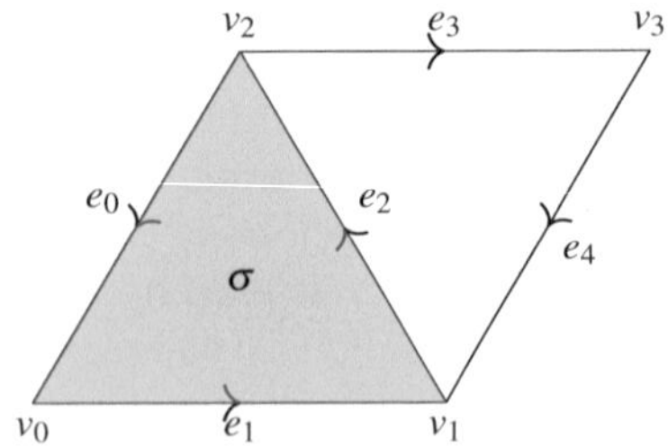

Fig. 1 A simplicial complex

One issue is that the usual algorithms for computing a homology generator for $H_q(X)$ gives no guarantees about the optimality of the representatives z_i in the computed homology generator. In particular, inside a given homology class $h = [z]$ there may be a representative $\tilde{z} \sim z$ that fulfills some optimality criterion. Alternatively, given a homology generator $G = \{z_1, \dots, z_m\}$, there may be a different one $G' = \{\tilde{z}_1, \dots, \tilde{z}_m\}$ satisfying some optimality criterion. The former is an optimal homologous cycle problem, the latter is an optimal homology generator problem.

For example, given the simplicial complex in Fig. 1, one can compute that $H_1(X) = \mathbb{Z}[c_1] = \mathbb{Z}[c_2]$ where $c_1 = e_3 + e_4 - e_1 - e_0$ and $c_2 = e_2 + e_3 + e_4$. The hole represented by c_2 can also be represented by c_1. How do we obtain "good" representatives for the computed homology classes? In this small example, the answer is clear by visual inspection. But in general, we use the linear programming method which we discuss in the next section.

2.2 Persistent Homology

Here, we give a short review of persistent homology [8]. Instead of a single simplicial complex, we consider a filtration, a nested sequence of simplicial complexes

$$\mathbb{X} : X_1 \subset X_2 \subset \cdots \subset X_n$$

as input.

Because of technical conditions, instead of using $\mathbb{Z}$ as the coefficient ring, we need to use a field. Here, we use the field $\mathbb{Q}$. Thus, the chain groups and homology groups are vector spaces over $\mathbb{Q}$. An alternative would be to use the finite field $\mathbb{Z}_2$, but for the purpose of using existing linear programming packages we use $\mathbb{Q}$.

We take the homology group of each simplicial complex in the above sequence. Moreover, the inclusions induce maps between the homology groups in the following way. Clearly, we have an inclusion $\iota : C_q(X_{i-1}) \to C_q(X_i)$ and then inclusions $Z_q(X_{i-1}) \subset Z_q(X_i)$, $B_q(X_{i-1}) \subset B_q(X_i)$. Thus, define $\iota^*_{i-1,i} : H_q(X_{i-1}) \to H_q(X_i)$ by mapping the homology class $z + B_q(X_{i-1})$ to $z + B_q(X_i)$. We have

$$H_q(\mathbb{X}) : H_q(X_1) \xrightarrow{\iota^*_{1,2}} H_q(X_2) \xrightarrow{\iota^*_{2,3}} \cdots \xrightarrow{\iota^*_{n-1,n}} H_q(X_n). \tag{1}$$

By appropriate transformation, we can write $H_q(\mathbb{X})$ uniquely as a direct sum of the interval indecomposables. In terms of algebra, this means we have an isomorphism

$$H_q(\mathbb{X}) \cong \bigoplus_{1 \le i \le j \le n} \mathbb{I}[i, j]^{m_{i,j}}, \tag{2}$$

where each interval $\mathbb{I}[i, j]$ appears with some multiplicity $m_{i,j}$. The term $\mathbb{I}[i, j]$ is given by the diagram

$$0 \to \cdots \to 0 \to \mathbb{Q} \xrightarrow{1} \cdots \xrightarrow{1} \mathbb{Q} \to 0 \to \cdots \to 0$$

with nonzeros from indices i to j.

In practical terms, Eq. (2) tells us that we can decompose $H_q(\mathbb{X})$ into intervals, each interval $\mathbb{I}[i, j]$ representing a topological feature born at some index i and lasting until index j. In other words, at index $j+1$ and onwards, the feature no longer exists.

In this notation, a feature represented by an interval $[i, j]$ is present from i to j inclusive. For the persistence algorithm, we denote the so-called persistence intervals by (b, d), where $b = i$ is the birth index and $d = j+1$ is the death index. At index d, the feature dies off. If $j = n$, then the feature represented by $\mathbb{I}[i, j]$ does not disappear even at the end of the filtration and we equate the past-the-end index $d = j + 1$ to ∞.

The information in Eq. (2) can be summarized, up to isomorphism, using the so-called *persistence diagram*. Given $H_q(\mathbb{X})$, its persistence diagram is a multiset of intervals (b, d) with multiplicity $m_{i,j} = m_{b,d-1}$ as determined by Eq. (2). We plot the persistence diagram as a multiset of points. Clearly, all points in the persistence diagram are above the diagonal. The points near the diagonal have short lifespan $d - b$, while those far from the diagonal have long lifespans, and can be interpreted as being topologically robust, or significant.

2.3 Linear Optimization

A vector x is said to be nonnegative if all of its entries are nonnegative. We write $x \geq 0$ in such a case. Similarly, x is said to be integral if all of its entries are integers.

Let b, c be vectors, and A be a matrix of appropriate size. A (minimization) linear optimization problem is

$$\begin{aligned} &\text{minimize} \quad c'x \\ &\text{subject to} \quad \begin{cases} Ax = b, \\ x \geq 0. \end{cases} \end{aligned} \tag{3}$$

Here, c' is the transpose of c, and the function $f(x) = c'x$ is called the objective function of this problem. The constraints in this problem are the equations $Ax = b$ and $x \geq 0$; the set of x that satisfy the constraints is called the feasible region. We call the matrix A the constraint matrix of this problem. An x in the feasible region that gives the minimum value of $c'x$, if it exists, is called an optimal solution to this problem, with optimum value $c'x$.

If we add the condition that x be integral, then the problem becomes an integer linear programming problem. In general, this problem is known to be NP-hard.

3 Problem Formulation

Before considering optimal cycles in persistent homology in Sect. 3.3, we first review some more background. In Sect. 3.1, we discuss the optimal homologous cycle problem posed in [5] together with some of their results. In Sect. 3.2, we recall the extension to the multiple cycle case from [10]. In both Sects. 3.1 and 3.2, we cast the problems as integer programming problems, while in Sect. 3.3 we drop the integrality requirement.

For each q, we take as standard basis for $C_q(X)$ the set of q-dimensional simplices $\{\sigma_1, \dots, \sigma_N\}$. Then, we write $x = \sum_{i=1}^{N} x_i \sigma_i \in C_q(X)$ as a vector $[x_1, \dots, x_N]^T$. Given these bases, we can write each ∂_q in matrix form with respect to the bases for $C_q(X)$ and $C_{q-1}(X)$. Furthermore, in this basis, we identify x with the vector representing x and ∂_q with its matrix form.

3.1 Single Cycle Case

We recall the optimal homologous cycle problem as posed in [5]. Given a cycle $z \in Z_q(X)$, consider the problem:

$$
\begin{aligned}
&\text{minimize} \quad ||x||_1 \\
&\text{subject to} \quad \begin{cases} x - \partial_{q+1} y = z, \\ x, y \text{ integral.} \end{cases}
\end{aligned}
\tag{4}
$$

Here, we try to find a chain $\tilde{z}$ homologous to z that has minimum 1-norm among all chains homologous to z. Recall that the 1-norm is defined as $|| \sum_{i=1}^{N} x_i \sigma_i ||_1 = \sum_{i=1}^{N} |x_i|$. Clearly, an optimal solution exists for Problem (4). Given an optimal solution $x = \tilde{z}$ and $y = \tilde{y}$ to this problem, we call $\tilde{z}$ an *optimal cycle* homologous to z.

Alternatively, we can formulate the problem of finding a homologous cycle with minimum number of simplices. That is, we define the 0-norm $|| \sum_{i=1}^{N} x_i \sigma_i ||_0 = |\{x_i \neq 0\}|$ on $C_q(X)$. Then, we get $\tilde{z} \in C_q(X)$ as a minimizer for $||x||_0$ among all $x \sim z$. Moreover, to facilitate interpretation of $\tilde{z}$ geometrically, we require that $\tilde{z} = \sum_\sigma n_\sigma \sigma$ satisfies $n_\sigma \in \{-1, 0, 1\}$ for all σ. Restricted to $\{-1, 0, 1\}$-vectors, it is clear that $||x||_0 = ||x||_1$ for all feasible x. We have

$$
\begin{aligned}
&\text{minimize} \quad ||x||_1 \\
&\text{subject to} \quad \begin{cases} x - \partial_{q+1} y = z, \\ x \text{ is a } \{-1, 0, 1\}\text{-vector}, \ \ y \text{ is integral.} \end{cases}
\end{aligned}
\tag{5}
$$

Note that Problem (5) does not always have a solution. An easy example is given by with $z = 2e_2 + 2e_3 + 2e_4 \in Z_1(X)$ of the simplicial complex in Fig. 1. If z

itself is a $\{-1, 0, 1\}$-vector, then an optimal solution exists for Problem (5). See [5] for more details. Where possible, we prefer to have an optimal solution that is a $\{-1, 0, 1\}$-vector for the geometric interpretation.

We write x as $x^+ - x^-$, with $x^+, x^- \geq 0$ corresponding to the positive and negative parts of x. We can rewrite the problem as

$$\begin{aligned} &\text{minimize} \quad ||x||_1 = \sum_{i=1}^{N} (x_i^+ + x_i^-) \\ &\text{subject to} \quad \begin{cases} (x^+ - x^-) - \partial_{q+1} y = z, \\ x^+, x^- \text{ are } \{0, 1\}\text{-vectors}, \ y \text{ integral}, \end{cases} \end{aligned}$$

where x_i^+, x_i^- are the entries of the vectors x^+, x^-, respectively. Given an optimal solution $x = x^+ - x^- = \tilde{z}$, and $y = \tilde{y}$ to this problem, we also call $\tilde{z}$ an optimal cycle homologous to z. Note that by the constraints imposed, we have $\partial_q \tilde{z} = \partial_q(z + \partial_{q+1} y) = 0$ by the fact that z is a cycle and that $\partial_q \partial_{q+1} = 0$.

We also warn that the optimal solution for any of the problems above may not be unique. It is easy to construct examples where this is the case.

Recall that a matrix is said to be totally unimodular if the determinant of each of its square submatrices is -1, 0, or 1. A basic fact in integral programming is the following. See [16], for example.

Lemma 1 *Let A be totally unimodular, b be an integral vector, and c' be some row vector. Then, the integer linear optimization problem*

$$\begin{aligned} &\textit{minimize} \quad c'x \\ &\textit{subject to} \quad \begin{cases} Ax = b \\ x \geq 0, \textit{ integral} \end{cases} \end{aligned} \tag{6}$$

can be solved in time polynomial in the size of A.

This can be accomplished by dropping the integrality requirements, and solving the problem as a linear programming problem over the reals, which is known to have a polynomial time algorithm. Then, total unimodularity guarantees that the optimal solution will be integral.

We need to first write both problems into standard form as in Problem (6). For example, Problem (4) can be written as

$$\begin{aligned} &\text{minimize} \quad \sum_{i=1}^{N} (x_i^+ + x_i^-) \\ &\text{subject to} \quad \begin{cases} x^+ - x^- - \partial_{q+1}(y^+ - y^-) = z, \\ x^+, x^-, y^+, y^- \geq 0, \ \text{integral} \end{cases} \end{aligned}$$

and thus has constraint matrix $A = [\, I \ -I \ -\partial_{q+1} \ \partial_{q+1} \,]$. Problem (5) can also be written into such a form, where we have to incorporate the constraint that x be a $\{-1, 0, 1\}$-vector into the constraint matrix. Then, it can be seen that if ∂_{p+1} is totally unimodular, so are the constraint matrices corresponding to Problems (4) and (5).

For a fixed $q \geq 0$, Dey et al. [5] gives some conditions for the matrix ∂_{q+1} to be totally unimodular. For example, if X is a finite simplicial complex triangulating a compact, $(q+1)$-dimensional manifold or if X is a finite simplicial complex embedded in $\mathbb{R}^{q+1}$, then ∂_{q+1} is totally unimodular.

In the next subsection, where we consider the multiple cycle case, the addition of extra columns determined by other cycles to the constraint matrix may destroy total unimodularity. Nevertheless, in this work, we ignore the difficulty of NP-hardness and formulate our problem as below.

3.2 Multiple Cycle Case

In general, the optimal solution obtained by solving the optimal homologous cycle problem may not be enough for some purposes. For example, consider Fig. 2. In this case, the cycle z represented by the dashed loop optimizes to $\tilde{z}$ represented by the dotted loop, if we solve Problem (5). Intuitively speaking, the optimal cycle homologous to z is stuck around the two holes. We would like to be able to detect both holes individually. We are no longer seeking the optimum inside the homology class of z, but considering generators of homology.

We modify the problem as follows. Let $\{g_1, \dots, g_m\}$ be a set of cycles. We specify how we set these cycles later. For now, consider the problem

$$\begin{aligned} &\text{minimize} \quad ||x||_1 \\ &\text{subject to} \quad \begin{cases} x - \partial_{q+1} y + \sum_{j=1}^{m} a_j g_j = z, \\ x,\ y, \text{ and } a \text{ are integral,} \end{cases} \end{aligned} \tag{7}$$

and let $P(z; g_1, \dots, g_m)$ be the set of optimal solutions projected to the x variable of this problem. That is, an optimal solution is given by a triple $x = \tilde{z}$, $y = \tilde{y}$, $a = \tilde{a}$, and we let $P(z; g_1, \dots, g_m)$ consist of the vectors $\tilde{z}$.

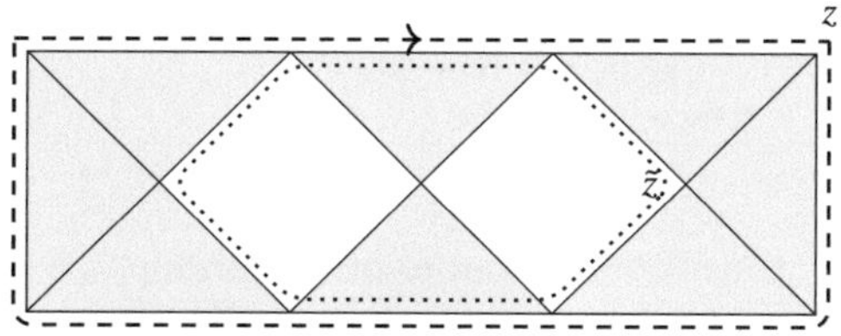

Fig. 2 Multiple nonhomologous cycles

Of course, we can linearize this problem as before:

$$\begin{array}{ll} \text{minimize} & \sum_{i=1}^{N}(x_i^+ + x_i^-) \\ \text{subject to} & \begin{cases} (x^+ - x^-) - \partial_{q+1} y + \sum_{j=1}^{m} a_j g_j = z, \\ x^+, x^- \geq 0,\ x^+, x^-,\ y, \text{ and } a \text{ are integral.} \end{cases} \end{array}$$

For simplicity, in this problem we have dropped the requirement that x be a $\{-1, 0, 1\}$-vector.

The extra cycles $g_1, \ldots, g_j$ are cycles that we wish to "factor out" in optimizing the current cycle z. We call these extra cycles the *relative cycles* in this problem. The summand $\sum_{j=1}^{m} a_j g_j$ can be understood in the following way. Let

$$\partial'_{q+1} = [\, \partial_{q+1}\ g_1\ \cdots\ g_m \,]$$

be the boundary matrix with extra columns $g_1, \ldots, g_m$ appended. Geometrically speaking, we are inserting additional cells $\tau_1, \ldots, \tau_m$ into X_{q+1} with $\partial'_{q+1}\tau_j = g_j$ for $j = 1, \ldots, m$. Each cell τ_j covers the topological feature represented by g_j, allowing us to minimize z going across g_j. Note that the resulting structure $X' = X \cup \{\tau_j\}$ is in general no longer a simplicial complex, but may be described as a cell complex. Moreover, even if ∂_{p+1} were totally unimodular, the addition of these new columns may cause ∂'_{q+1} to not be totally unimodular. See [10] for details and examples.

We again emphasize that a solution $\tilde{z}$ to Problem (7) may not be homologous to z. Here, we are no longer optimizing inside a single homology class. Suppose that we have a homology generator $\{z_1, \ldots, z_n\}$ for $H_q(X)$. Then, replacing any z_j by an optimal solution

$$\tilde{z}_j \in P(z_j; z_1, \ldots, z_{j-1}, z_{j+1}, \ldots, z_n)$$

gives a new generating set $[z_1], \ldots, [\tilde{z}_j], \ldots, [z_n]$ for $H_q(X)$. This follows directly from the fact that $[\tilde{z}_j] = [z_j] + \sum_{i \neq j} a_i [z_i]$ due to the constraints. We perform the procedure OPTIMIZE_CYCLES.

Require: $\{z_1, \ldots, z_n\}$ homology generator for $H_q(X)$
  procedure OPTIMIZE_CYCLES($\{z_1, \ldots, z_n\}$)
    for $j = 1, \ldots, n$ **do**
      Choose a $\tilde{z}_j$ from $P(z_j; \tilde{z}_1, \ldots, \tilde{z}_{j-1}, z_{j+1}, \ldots, z_n)$
    return $\{\tilde{z}_1, \ldots, \tilde{z}_n\}$

Clearly the output to this algorithm is a new homology generator for $H_q(X)$. If we add the assumption that $H_q(X)$ is free, then we can replace the term homology generators by homology bases in the above discussion.

For example, let us apply OPTIMIZE_CYCLES to the setting given in Fig. 2. Let z_1 be the outer dashed cycle, and z_2 be the left cycle. Then, $\{[z_1], [z_2]\}$ generates

$H_1(X)$. In the first iteration, we solve $P(z_1; z_2)$:

$$\text{minimize} \quad ||x||_1$$
$$\text{subject to} \quad \begin{cases} x - \partial_{q+1} y + a z_2 = z_1, \\ x, y, \text{ and } a \text{ are integral.} \end{cases}$$

We get $\tilde{z}_1$, the cycle on the right representing the right hole. Next, we solve $P(z_2; \tilde{z}_1)$, giving us $\tilde{z}_2 = z_2$ itself as an optimal solution.

3.3 Optimal Cycles in Persistent Homology

First, let us recall the classical algorithm for persistent homology. To facilitate computations, we assume that the filtration

$$\emptyset = X_0 \subset X_1 \subset X_2 \subset \cdots \subset X_N$$

starts from the empty set, and at each index, only one new simplex is added. That is, we assume that $X_j \backslash X_{j-1} = \{\sigma_j\}$. With this assumption, each simplex σ_j causes either the birth or death of a homology class. Also, to each simplex in X_N is associated a unique birth time j.

We denote the boundary of the jth simplex σ_j by $A_j \doteq \partial \sigma_j$. We write this boundary as a vector in the standard basis of all simplices in X, ordered according to the indexing in the filtration. We denote the ith entry of A_j by $A_j(i)$. That is, $A_j(i)$ is the coefficient of σ_i in $\partial \sigma_j$. For any column A_j, we define its pivot row $\text{PIVOT}(A_j)$ to be the largest row index i with nonzero entry $A_j(i)$.

---

procedure COMPUTE_PERSISTENCE(X)
 Initialize $g_j = \sigma_j$ for $j = 1, \ldots, N$.
 for $j = 1, \ldots, N$ **do**
  while $A_j \neq 0$ and $l(\text{PIVOT}(A_j)) \neq 0$ **do**
   $p \leftarrow \text{PIVOT}(A_j)$
   $r \leftarrow -\frac{A_j(p)}{A_{l(p)}(p)}$
   $A_j \leftarrow A_j + r \cdot A_{l(p)}$
   $g_j \leftarrow g_j + r \cdot g_{l(p)}$
  if $A_j \neq 0$ **then**
   $i \leftarrow \text{PIVOT}(A_j)$
   $l(i) \leftarrow j$
   Insert (i, j) into Pairs.
 for all i such that $i \neq b, d$ for all (b, d) in Pairs **do**
  Insert (i, ∞) into Pairs.
 return Pairs

The algorithm COMPUTE_PERSISTENCE computes the persistent intervals of the input filtration. Inside the loop over all simplices, there are two main parts. First is a reduction step given as a while loop where we reduce the current column A_j by adding multiples of other columns to it. We perform only column operations from left to right, due to the filtration structure of the input. Next, we check whether the reduction step zeroes out the current column.

Relative to the current state of the matrix A, the lookup table l records the index of the reduced column with pivot entry in a given row. Necessarily, we only record the indices of the columns to the left of the current column. In particular, $l(p)$ stores the index of the reduced column with pivot in row p. In the reduction step, suppose column A_j currently has pivot at row p. Then, column $A_{l(p)}$, which is a column to the left of j, has a nonzero in row p. We thus add a multiple of $A_{l(p)}$ to A_j to zero out $A_j(p)$.

If A_j is reduced to a zero column, we have the birth of a cycle g_j. Otherwise, it represents the death of the cycle that was born at $i = \text{PIVOT}(A_j)$. This gives us a persistence pair (i, j). We record $l(i) = j$, since column j has a nonzero at row i.

We store the basis changes in the chains g_j. At the end of the algorithm, every index i with reduced column $A_i = 0$ will either be paired to some j, or left unpaired. We denote

$$d(i) = \begin{cases} j & \text{if } (i, j) \text{ pair extracted}, \\ \infty & \text{if } i \text{ left unpaired}, \end{cases}$$

for the death index of the feature born at index i. Then, after running the algorithm, we let

$$\mathscr{L}_q(k) = \{i \,|\, i \leq k, A_i = 0, \dim \sigma_i = q, d(i) > k\}.$$

These indices correspond to cycles g_i of dimension q that have not become part of the boundary in X_k. It can be shown that

$$\{[g_i] \,|\, i \in \mathscr{L}_q(k)\}$$

forms a basis for $H_q(X_k)$. This follows from the standard arguments concerning persistent homology.

For further details of the persistence algorithm and proofs of these statements, we refer the reader to [8]. In this work, we augment the algorithm to compute optimal cycles. Essentially, whenever a cycle is born, we optimize it relative to the current state of the simplicial complex in the filtration.

For simplicity, we restrict our attention to cycles in some target dimension q. Of course, by keeping separate optimization problems and boundary matrices for each dimension, the algorithm can be extended to compute optimal cycles for more than one dimension at a time.

---

procedure PERSISTENCE_OPTIMAL_CYCLES(X,q)
  Initialize l a lookup map; B, an empty matrix.
  Initialize $g_j = \sigma_j$, $j = 1, \ldots, N$.
  Initialize empty optimization problem.
  for $j = 1, \ldots, N$ **do**
    if $\dim \sigma_j = q$ **then**
      Update optimization problem with new simplex σ_j.
    while $A_j \neq 0$ and $l(\text{PIVOT}(A_j)) \neq 0$ **do**
      $p \leftarrow \text{PIVOT}(A_j)$
      $r \leftarrow -\frac{A_j(p)}{A_{l(p)}(p)}$
      $A_j \leftarrow A_j + r \cdot A_{l(p)}$
      $g_j \leftarrow g_j + r \cdot g_{l(p)}$
    if $A_j = 0$ and $\dim \sigma_j = q$ **then**
      $\tilde{z}_j = \text{OPTIMIZE\_CYCLE}(g_j)$.
      Update optimization problem with $\tilde{z}_j$, a relative column
    if $A_j \neq 0$ **then**
      $i \leftarrow \text{PIVOT}(A_j)$
      $l(i) \leftarrow j$
      Insert (i, j) into Pairs.
      if $\dim \sigma_j = q + 1$ **then**
        Append A_j to matrix B (as a boundary column).
        Remove $\tilde{z}_i$ from the optimization problem.
  for all i such that $i \neq b, d$ for all (b, d) in Pairs **do**
    Insert (i, ∞) into Pairs.

Every time a simplex σ_j of dimension q is encountered, we update the optimization problem with a new variable and a new constraint corresponding to it. In Problem (8), the variable x increases in size by one and constraints are padded with zero entries.

---

procedure OPTIMIZE_CYCLE($z_j = g_j$)
  Find an optimal solution $\tilde{z}_j$ to

$$\begin{aligned} &\text{minimize} \quad ||x||_1 \\ &\text{subject to} \quad \left\{ x + By + \sum_{i \in \mathscr{L}_q(j), i<j} a_i \tilde{z}_i = z_j. \right. \end{aligned} \tag{8}$$

  return $\tilde{z}_j$

When a zero column $A_j = 0$ is created, this signals the birth of a new cycle. If it is in dimension q, then we solve the optimization problem. We take g_j, the cycle just born, as z_j, the right hand side in the constraint of the optimization Problem (8). Once we obtain an optimized cycle $\tilde{z}_j$, we append this result back into the optimization problem as a variable representing the relative cycle.

If we do not get a zero column after the reduction step, then as before this signals the creation of a birth-death persistence pair (i, j), where $i = \text{PIVOT}(A_j)$. If $\dim \sigma_j = q+1$, we add the column to the optimization problem as a boundary column. This is a new variable, so that the size of y in Problem (8) increases by one, and the column A_j is appended to the end of B.

This also means that the cycle $\tilde{z}_i$ of dimension q becomes part of the boundary, where i is the birth index paired to j. Accordingly, the relative cycle $\tilde{z}_i$ is no longer needed in the optimization problem and we may erase the corresponding variable.

The procedure OPTIMIZE_CYCLE solves a version of Problem (7), with B serving as the current state of the reduced columns of the boundary matrix restricted to simplices of dimension $q+1$, and the already optimized cycles $\tilde{z}_i, i < j$ that remain in the problem serving as the relative cycles.

Since we are computing persistence with $\mathbb{Q}$-coefficients, it suffices to perform linear optimization. Of course, we can impose additional conditions such as integrality or requiring the solution be a $\{-1, 0, 1\}$-vector, but depending on the form of g_j such additional constraints may make the problem infeasible.

For output, every persistence pair (i, j) or (i, ∞) with $\dim(\sigma_i) = q$ is associated to an optimal cycle representative $\tilde{z}_i$. Recall that

$$\{[g_i] | i \in \mathscr{L}_q(k)\}$$

forms a basis for $H_q(X_k)$. We now show that we can replace g_i by $\tilde{z}_i$ in the previous statement. Note first that the homology class of $\tilde{z}_i$ in $H_q(X_k)$ satisfies

$$[\tilde{z}_i] = [g_i] + \sum_{h \in \mathscr{L}_q(i), h<i} a_h [\tilde{z}_h]$$

by the given constraint in Problem (8) and since the image of the matrix B is simply $\operatorname{im} \partial_{q+1}|_{X_k}$. Then, using similar equations for $[\tilde{z}_h]$, we can ultimately write

$$[\tilde{z}_i] = [g_i] + \sum_{h<i, A_h=0, \dim \sigma_h = q} c_h [g_h]. \tag{9}$$

In expanding each $[\tilde{z}_h] = [g_h] + \sum_{h' \in \mathscr{L}_q(h), h'<h} a'_{h'} [\tilde{z}_{h'}]$, it is not certain that each index h' so obtained corresponds to a cycle $g_{h'}$ still living at index k. Nevertheless, for those h' that have $d(h') \leq k$, $[g_{h'}] = 0$ since they die off in or before $H_q(X_k)$. For those indices, $[g_h]$ merges into the boundary and thus we may remove those indices from the summation. We obtain

$$[\tilde{z}_i] = [g_i] + \sum_{h \in \mathscr{L}_q(k), h<i} c_h [g_h] \tag{10}$$

from Eq. (9). Thus, replacing any (or all) of the $[g_i]$ by $[\tilde{z}_i]$, we still have a basis for $H_q(X_k)$. This can be easily seen from the fact that the transformation induced by Eq. (10) is invertible. Written as a matrix, the requirement that $h < i$ implies that the transformation is triangular with only 1's in the diagonal.

Depending on application, one may not need to compute OPTIMIZE_CYCLE(g_j) for all j with $\sigma_j = q$. For example, suppose that we have a particular range of birth indices J, or even a particular birth index $J = \{j'\}$ in which we are interested. Then, we could instead compute only OPTIMIZE_CYCLE$(g_{j'})$ for $j' \in J$ and for the other indices simply update the optimization problem with g_j. That is, by replacing the appropriate part of PERSISTENCE_OPTIMAL_CYCLES by the following code fragment:

```
if A_j = 0 and dim(σ_j) = q then
    if j ∈ J then
        z̃_j = OPTIMIZE_CYCLE(g_j).
        Update optimization problem with z̃_j, a relative column
    else
        Update optimization problem with g_j, a relative column
```

we can choose which indices to optimize. We can replace the constraint

$$x + By + \sum_{i \in \mathscr{L}_q(j), i<j} a_i \tilde{z}_i = z_j$$

in Problem (8) by

$$x + By + \sum_{i \in \mathscr{L}_q(j), i<j} a_i g_i = z_j.$$

In either formulation, the feasible regions projected to x are the same, by a similar argument as above. Thus, we can skip optimizing some of the cycles g_i and still solve the same problem as if we had optimized all of them.

4 Example

Given a set of points $\{x_i\}_{i=1}^M$ in $\mathbb{R}^N$, which we call a *point cloud*, together with a weight $w_i = r_i^2$ for every point, we construct the so-called weighted alpha complex [7]. The weighted alpha complex is a simplicial complex constructed over the point cloud, and is useful for modelling molecular structures. The weighted alpha complex is controlled by a parameter α, which can be interpreted as a scale parameter. This naturally results in a filtration of simplicial complexes. That is, for $\alpha < \alpha'$, we have $X_\alpha \subset X_{\alpha'}$. In fact, the weighted alpha complex has same homotopy type as the following model. Around each point, we define a ball $B_i(\alpha)$ with radius $r_i(\alpha) = \sqrt{r_i^2 + \alpha}$, and we are studying the topological features in the union of balls model $\bigcup_{i=1}^M B_i(\alpha)$.

Fig. 3 Dimension 1 persistence diagram of 2HHB. We note the interval with largest lifespan, at the point (−0.59572, 18.5573). Also, we have encircled four points with the top four largest lifespans

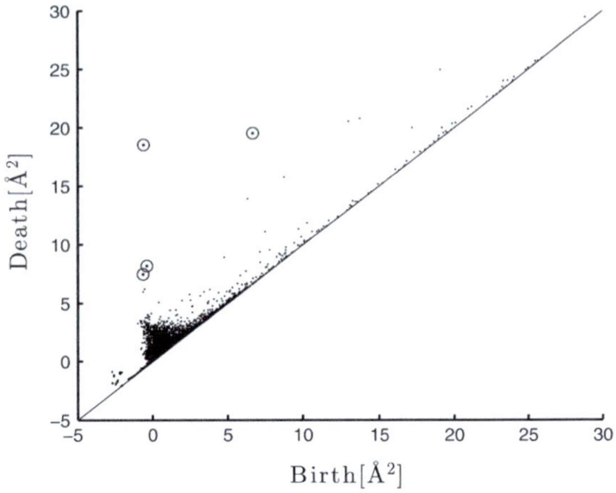

We take as input data the atomic structure of the protein human deoxyhemoglobin (PDB ID:@ 2HHB [11]) from the Protein Data Bank [1] at http://www.rcsb.org and construct its weighted alpha complex using CGAL [2]. We use the square of the atomic radii as the weights. We apply the algorithm PERSISTENCE_OPTIMAL_CYCLES to the weighted alpha complex filtration. We use the software CPLEX version 12.6 [12] to implement and solve the linear programming problems.

We plot the persistence diagram with $q = 1$ in Fig. 3. For every interval in the persistence diagram, we compute a corresponding optimal cycle, but we do not display all of them. We consider only the optimal cycles of the intervals with the top four largest lifespans.

Fig. 4 An optimal cycle corresponding to the persistence interval with largest lifespan

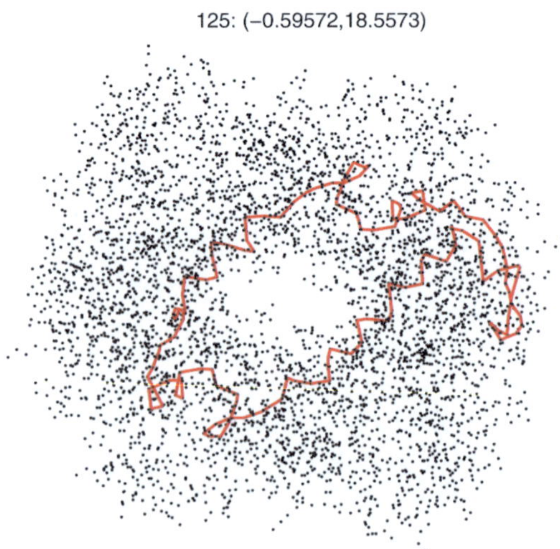

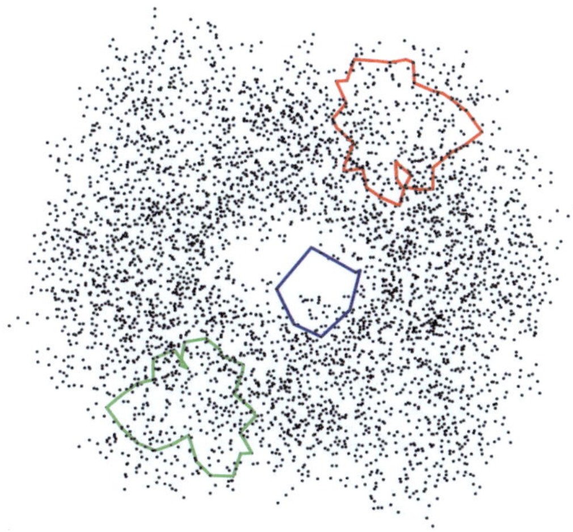

Fig. 5 Optimal cycles corresponding to the intervals with 2nd to 4th largest lifespans. In order of decreasing lifespan, *blue* has length 8 and interval $(6.6614, 19.5345)$, *green* has length 28 and interval $(-0.39261, 8.1848)$, and *red* has length 34 and interval $(-0.63727, 7.4801)$

Sorting all of the persistence intervals by length of lifespan $d - b$, we have the interval $(-0.59572, 18.5573)$ with largest lifespan. We plot this optimal cycle in Fig. 4 together with the point cloud. This optimal cycle winds around the main hole. We also display in Fig. 5 the optimal cycles corresponding to the intervals with the 2nd to the 4th largest lifespans.

Acknowledgments The authors would like to thank Genki Kusano and Hiroshi Takeuchi for interesting discussions, especially from the side of applications, and Prof. Hayato Waki for the chance to give this talk in the workshop "Optimization in the Real World—Towards Solving Real-World Optimization Problems".

References

1. Berman, H.M., Westbrook, J., Feng, Z., Gilliland, G., Bhat, T.N., Weissig, H., Shindyalov, I.N., Bourne, P.E.: The Protein Data Bank. Nucleic Acids Res. **28**(1), 235–242 (2000)
2. CGAL, Computational Geometry Algorithms Library: http://www.cgal.org
3. Chen, C., Freedman, D.: Hardness results for homology localization. Discrete Comput. Geom. **45**(3), 425–448 (2011)
4. Chen, C., Freedman, D.: Measuring and computing natural generators for homology groups. Comput. Geom. **43**(2), 169–181 (2010)
5. Dey, T.K., Hirani, A.N., Krishnamoorthy, B.: Optimal homologous cycles, total unimodularity, and linear programming. SIAM J. Comput. **40**(4), 1026–1044 (2011)
6. Dey, T.K., Sun, J., Wang, Y.: Approximating cycles in a shortest basis of the first homology group from point data. Inverse Prob. **27**(12) (2011)
7. Edelsbrunner, H.: Weighted alpha shapes. Technical report, Department of Computer Science, University of Illinois at Urbana-Champaign (1992)
8. Edelsbrunner, H., Letscher, D., Zomorodian, A.: Topological persistence and simplification. Discrete Comput. Geom. **28**(4), 511–533 (2002)
9. Erickson, J., Whittlesey, K.: Greedy optimal homotopy and homology generators. In: Proceedings of the ACM-SIAM Symposium on Discrete Algorithms, pp. 1038–1046. SIAM (2005)

10. Escolar, E.G., Yasuaki H.: Computing optimal cycles of homology groups. In: Nishii, R., et al. (eds.) A Mathematical Approach to Research Problems of Science and Technology, pp. 101–118. Springer, Japan (2014)
11. Fermi, G., Perutz, M.F., Shaanan, B., Fourme, R.: The crystal structure of human deoxyhaemoglobin at 1.74 Å resolution. J. Mol. Biol. **175**(2), 159–174 (1984)
12. International Business Machines Corp.: IBM ILOG CPLEX Optimization Studio. http://www-03.ibm.com/software/products/en/ibmilogcpleoptistud/
13. Kaczynski, T., Mischaikow, K., Mrozek, M.: Computational Homology. Springer, New York (2004)
14. Munkres, J.: Elements of Algebraic Topology. The Benjamin/Cummings Publishing Company Inc, Menlo Park, California (1984)
15. Nakamura, T., Hiraoka, Y., Hirata, A., Escolar, E.G., Matsue, K., Nishiura, Y.: Description of Medium-Range Order in Amorphous Structures by Persistent Homology. arxiv.org/abs/1501.03611
16. Salkin, H.M., Mathur, K.: Foundations of Integer Programming. North Holland, New York (1989)
17. Zomorodian, A., Carlsson, G.: Computing persistent homology. Discrete Comput. Geom. **33**(2), 249–274 (2005)

Optimal Battery Control for Smart Grid Nodes

Andreas Draegert, Andreas Eisenblätter, Inken Gamrath and Axel Werner

Abstract Energy storages can be of great value when added to power grids. They introduce the possibility to store and release energy whenever this is favorable. This is particularly relevant, for example, if power supply is volatile (as is the case with renewable energy) and the network is small (so that there are few other nodes that might balance fluctuations in consumption or production). We present models and methods from mathematical optimization for computing an optimized storage schedule for this purpose. We look at alternative optimization objectives, such as smallest possible peak load, lowest energy costs, or the closest approximation of a prescribed load curve. The optimization needs to respect general operational and economic constraints as well as limitations in the use of storage, which are imposed by the chosen storage technology. We therefore introduce alternative approaches for modeling the non-linear properties of energy storages and study their impact on the efficiency of the optimization process. Finally, we present a computational study with batteries as storage devices. We use this to highlight the trade-off between solution quality and computational tractability. A version of the model for the purpose of leveling peaks and instabilities has been implemented into a control system for an office-building smart grid scenario.

Keywords Energy storage · Optimization models · Mathematical models · Nonlinear systems

---

A. Draegert · A. Eisenblätter
atesio GmbH, Bundesallee 89, 12161 Berlin, Germany
e-mail: draegert@atesio.de

A. Eisenblätter
e-mail: eisenblaetter@atesio.de

I. Gamrath (✉) · A. Werner
Department Mathematical Optimization, Zuse Institute Berlin, Takustraße 7, 14195 Berlin, Germany
e-mail: inken.gamrath@zib.de

A. Werner
e-mail: werner@zib.de

K. Fujisawa et al. (eds.), *Optimization in the Real World*,
Mathematics for Industry 13, DOI 10.1007/978-4-431-55420-2_6

1 Introduction

The necessity of storing energy increases with the growing influence of renewable energies. This is particularly the case for consumers in a power grid who change their role at times and produce energy. For them, the significance of storing energy lies in saving money by partly decoupling from the spot market and the possibility of decreasing the system usage fee (transmission fee, or wheeling charge), or—more generally—by implementing a beneficial target profile for their power in-/output from and to the grid. Besides pumped hydroelectricity systems and compressed air storage, the storage of energy in batteries is an important technology used in practice; see, for instance, [1, 2].

In this paper, we present a general model that uses an abstract representation of the physical properties of the chosen storage technology and provides an optimized storage schedule, based on given energy prices, transmission fees depending on a peak transmission, or a given target purchase profile. We also describe—as a special case—how to use the capacity of batteries to flexibly buffer energy. Here, nonlinearities have to be mastered that originate from the physics of battery storage devices. For this, there are various approaches and the more precise the physical properties are modeled, the more computationally challenging are the resulting problems. To demonstrate this, we compare different approximations of the nonlinear discharge factor under various objectives, in terms of solution quality and run-time of the solving process. An implementation of the described approaches has been integrated in an operational environment to control the batteries at an ICT (information and communication technology) central office location; see http://www.desi-it2green.de for details.

Similar modeling approaches for optimal energy trading have been discussed in the literature, see [3–5]. These models, however, do not take detailed technical prerequisites into account. Similarly, various approaches modeling the management of residential electricity demands (see [6–8]) leave out the technical details. Besides this, literature provides—to the best of our knowledge—no comparable work that considers the computation of optimized storage schedules with a detailed view on the nonlinear technical constraints.

2 Modeling

We consider a node in a smart grid (usually a consumer, possibly a small energy producer) and assume an energy storage device, together with a controller, is available at the connection of the node to the power grid; see Fig. 1 for a schematic. Using this storage device, energy for consumption or from production at the node can be buffered. This can be done for a number of reasons, such as economical ones like market prices, or environmental ones like weather conditions that influence power generation. Buffering energy, however, involves losses (which depend on the storage

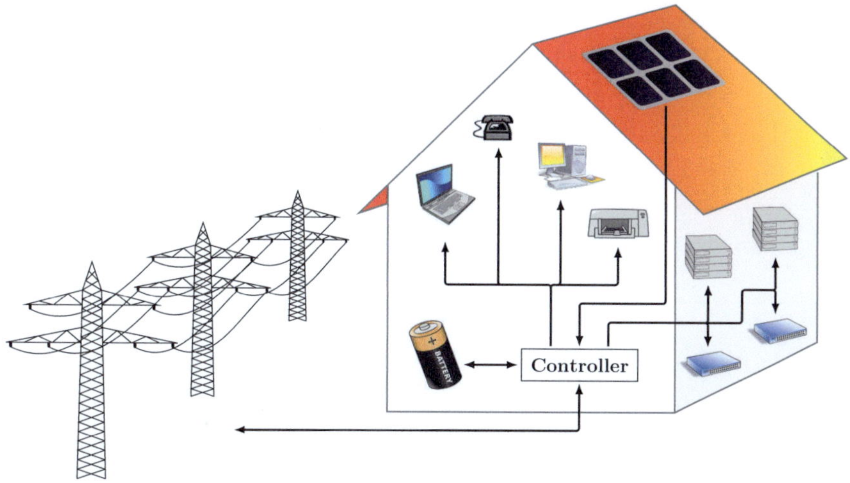

Fig. 1 General setting: Energy consumed or produced at a smart grid node can be buffered using a storage device (such as a *battery*) with a controller

technology used) and in deciding when the conditions for either storing or retrieving energy are best lies the main optimization potential.

We give a mathematical model for the problem of how to find an optimal controller schedule for the storage device, and we consider various objective functions that seem reasonable in this setting. Given energy prices, the cost factor for the transmission fee, or a target purchase profile for a certain future time period, as well as further input describing technological and planning properties, the model provides a schedule for storing and retrieving energy at the considered node for the said time period. To apply the model in a realistic scenario, it has to be adapted to a specific storage technology. We give an example for this in Sect. 3.

The model makes use of a discretization of the time scale. The optimization period $T = \{1, \ldots, t_{\max}\}$ comprises a finite number of time intervals of equal length. In our computations we use time steps of 15 min length—corresponding to the duration of an accounting interval—and a 24 h time horizon, so that we get $T = \{1, \ldots, 96\}$. Both assumptions are obviously not limitations of the approach and finer or rougher discretization may also be used. An important consequence of the fixed time discretization is that from the model's point of view power and energy are in fixed relation: The energy E stored or retrieved within a time step is directly proportional to the power P constantly applied over this time. The conversion factor is determined by the fixed length of time steps. We therefore widely speak of energy when—in a technical sense—energy and power would have to be distinguished.

An overview of the parameters, variables and further notation used in the modeling is given in Table 1.

Table 1 Model parameters and variables

Notation	Description
$T := \{1, \ldots, t_{\max}\}$	Set of time intervals
Parameters	
p_t, d_t	Expected energy produced/demanded in time step t
$S^{\text{init}}, S^{\min}, S^{\max}$	Initial and permissible minimal and maximal storage levels
$f^{\text{sto}}, f^{\text{ret}}, f^{\text{self}}$	Functions modeling various energy losses
$u_t^{\text{sto}}, u_t^{\text{ret}}, u_t^{\text{buy}}, u_t^{\text{sell}}$	Upper bounds on storing, retrieving, buying, and selling energy, respectively, in time step t
Variables	
$e_t^{\text{buy}}, e_t^{\text{sell}}$	Energy purchased/sold in time step t
$e_t^{\text{sto}}, e_t^{\text{ret}}$	Energy stored in/retrieved from storage in time step t
s_t	Storage level in time step t
x_t	Determines that energy is either stored or retrieved in time step t

2.1 Storage and Purchase Schedules

Table 2 presents the part of the model that defines feasibility: Constraints (1)–(3) address energy conservation and Constraints (4)–(5) define the correct storage level. In Sect. 2.2, we present a choice of objective functions, some of which can be used in combination.

Similar formulations have been described in the literature [3–5], where, however, none of the technical side constraints were taken into account.

Table 2 General model for computing a storage schedule

Energy conservation		
$e_t^{\text{buy}} + p_t + e_t^{\text{ret}} = e_t^{\text{sell}} + d_t + e_t^{\text{sto}}$	$\forall t \in T$	(1)
$e_t^{\text{sto}} - x_t u_t^{\text{sto}} \leq 0$	$\forall t \in T$	(2)
$e_t^{\text{ret}} - (1 - x_t) u_t^{\text{ret}} \leq 0$	$\forall t \in T$	(3)
Storage		
$s_0 = S^{\text{init}}$		(4)
$s_t = s_{t-1} f^{\text{self}}(s_{t-1}) + e_t^{\text{sto}} f^{\text{sto}}(e_t^{\text{sto}}, s_{t-1}) - e_t^{\text{ret}} f^{\text{ret}}(e_t^{\text{ret}}, s_{t-1})$	$\forall t \in T$	(5)
Variables		
$e_t^{\text{buy}} \in [0, u_t^{\text{buy}}]$	$\forall t \in T$	
$e_t^{\text{sell}} \in [0, u_t^{\text{sell}}]$	$\forall t \in T$	
$e_t^{\text{sto}} \in [0, u_t^{\text{sto}}]$	$\forall t \in T$	
$e_t^{\text{ret}} \in [0, u_t^{\text{ret}}]$	$\forall t \in T$	
$s_t \in [S^{\min}, S^{\max}]$	$\forall t \in T \cup \{0\}$	
$x_t \in \{0, 1\}$	$\forall t \in T$	

Energy Conservation

The continuous variables e_t^{buy} and e_t^{sell} announce the amount of energy purchased and sold, respectively, at the spot market at time interval t. The storage schedule is given by the continuous variables e_t^{sto} and e_t^{ret}, which denote the amount of stored and retrieved energy, respectively, in each time step. Upper bounds u_t^* on the purchased, sold, stored, or retrieved energy per time step may apply, for instance, to ensure that a given peak value of purchased energy is not exceeded or the energy to be stored or retrieved is within technically prescribed limits (cf. Sect. 3).

The first set of energy constraints (1) guarantees that all available energy—produced, purchased, and retrieved energy—is consumed, stored, or sold. By the second and third set (2) and (3) simultaneous storage and retrieval of energy in the same time interval is forbidden (which can be seen as a restriction imposed by the modeling).

Storage

To control the total amount of stored energy, we have continuous variables s_t for the storage level in each time step, which is allowed to vary in a given range from $S^{\min}$ to $S^{\max}$. It is set to an initial value for the beginning of the optimization period by constraint (4).

The last constraint set (5) describes the change in the storage level from one time step to the next. The functions f^{self}, f^{sto}, and f^{ret} determine the impact of losses that arise from self-discharging, storing, and retrieving energy over a time interval in dependence on the storage level s at the beginning of this time interval:

- $f^{\text{self}}(s) \in [0, 1]$ defines the fraction of energy remaining after losses due to self-discharging
- $f^{\text{sto}}(s, e^{\text{sto}}) \in [0, 1]$ describes the fraction of energy remaining after losses in the charging process (assuming a fixed charging scheme over the time interval)
- $f^{\text{ret}}(s, e^{\text{ret}}) \geq 1$ specifies the discharge factor; more precisely, the quotient of discharged energy, the amount by which the storage level is reduced, and retrieved energy, usable after discharging (assuming a constant discharging current over the time interval).

The nature of these functions depends on the used storage technology and has a decisive influence on the mathematical type of the model, as well as its solvability; see Sect. 3 for a more detailed discussion.

Miscellaneous

Further restrictions can be added to the model that allow the application of the optimization model in various scenarios.

When applying the optimization in a repetitive pattern, controlling the *final storage level*

$$s_{t_{\max}} \geq S^{\text{fin}} \tag{6}$$

at the end of the present time horizon may be desired. This allows to avoid schedules in which the stored energy is completely used up towards time $t_{\max}$. Similarly, the

Table 3 Additional model parameters and variables for further restrictions and objective functions

Notation	Description
t_{max}	Last time interval of the optimization horizon
Parameters	
S^{fin}	Final storage level
Objective coefficients	
$c_t^{\text{buy}}, c_t^{\text{sell}}$	Cost/revenue of trading energy in time step t
E^{peak}	Current peak transmission before actual optimization horizon
c^{peak}	Cost factor for transmission fee
$a_t^{\text{buy}}, a_t^{\text{sell}}$	Value for time step t in a desired purchase/selling profile
Variables	
y_t	Determines that energy is either purchased or sold at the spot market in time step t

battery level at the end of the optimization period might be required to reach a specific value for reasons such as planned maintenance.

Solutions that buy and sell energy during the same time interval may be undesired and forbidden:

$$e_t^{\text{buy}} - y_t u_t^{\text{buy}} \leq 0, \quad e_t^{\text{sell}} - (1 - y_t) u_t^{\text{sell}} \leq 0 \qquad \forall t \in T \tag{7}$$

where $y_t \in \{0, 1\}$ determines that we either buy or sell energy at the spot market. Note that, in case energy costs are minimized (see Sect. 2.2) and buying energy is more expensive than selling energy, these constraints are redundant for optimality.

Additional parameters for these restrictions as well as for the different objective functions are given in Table 3.

2.2 Objectives

The constraints given in Sect. 2.1 define a *feasible schedule* for the considered time horizon. Optimization itself consists now of finding a feasible schedule that is *optimal* with respect to a certain objective function. In the following, we list some objectives z that we considered. Every optimization problem can be formulated as a minimization problem, hence we are always looking for a feasible schedule that minimizes z.

Minimization of Energy Costs

Given spot market prices $c_t^{\text{buy}} \in \mathbb{R}$ and $c_t^{\text{sell}} \in \mathbb{R}$ for buying and selling energy at time t, respectively, the objective is to minimize the total energy cost

$$z(e_t^{\text{buy}}, e_t^{\text{sell}}) = \sum_{t \in T} \left(c_t^{\text{buy}} e_t^{\text{buy}} - c_t^{\text{sell}} e_t^{\text{sell}} \right). \tag{8}$$

Minimization of Peak Power

In some applications, the transmission fee (or wheeling charge) depends on the highest annual mean, which is defined as $4 \max e_t^{\text{buy}}$ (maximizing over all time steps t throughout 1 year). Then, minimizing the maximal load can be of interest:

$$z(e_t^{\text{buy}}) = \max_{t \in T} e_t^{\text{buy}}. \tag{9}$$

When the time horizon of an individual optimization run is shorter than 1 year, the objective for such runs can account for the increment of the peak value during the time horizon. The peak attained at earlier runs is propagated using a parameter E^{peak}. The objective then reads as

$$z(e_t^{\text{buy}}) = \max\left(0, \max_{t \in T}\left(e_t^{\text{buy}} - E^{\text{peak}}\right)\right). \tag{10}$$

Minimization of Combined Costs for Energy Purchase and Peak Consumption

The above objectives can be combined. Given the cost factor for the transmission fee $c^{\text{peak}} \in \mathbb{R}$, the objective to minimize the total cost[1] is

$$z(e_t^{\text{buy}}, e_t^{\text{sell}}) = c^{\text{peak}} \cdot \max_{t \in T} e_t^{\text{buy}} + \sum_{t \in T}\left(c_t^{\text{buy}} e_t^{\text{buy}} - c_t^{\text{sell}} e_t^{\text{sell}}\right). \tag{11}$$

As above, minimizing energy above a given "allowed" peak value might provide a better objective for consecutive optimization runs over a long time period.

Approximation of a Favored Purchase Profile

To minimize the deviation from a target profile for traded energy, given by values $a_t^{\text{buy}}, a_t^{\text{sell}} \geq 0$ for the considered time steps $t \in T$, we can formulate the objective

$$z(e_t^{\text{buy}}, e_t^{\text{sell}}) = \sum_{t \in T}\left(|e_t^{\text{buy}} - a_t^{\text{buy}}| + |e_t^{\text{sell}} - a_t^{\text{sell}}|\right). \tag{12}$$

3 Optimizing Battery Charging and Discharging Schedules

In Sect. 2, we presented a mathematical model for admissible charging/discharging and purchase schedules, as well as some examples for objective functions that have been used in practice. All details concerning physics and technology of the used storage device are hidden in the functions f^{self}, f^{sto}, and f^{ret}. Depending on those

[1] Scaling factors for the different terms may apply to account for the different time scales involved: whereas the optimization period is usually only 1 day, the transmission fees are computed based on the consumption in a full year, which makes it difficult to arrive at comparable cost values.

functions the nature of the model, as well as the available algorithmic tools for solving it, may vary largely.

This section gives an example how to make the general model precise for a specific situation. We consider one in which the storage device is a battery of total capacity Q. We emphasize that the specialization discussed in this section constitutes a rather rough version of a battery model and more realistic models can be implemented. Indeed, the level of detail of the technological model can be traded off against its solvability; we demonstrate this in the following by comparing the solutions and run-times using different possibilities for modeling the discharging factor f^{ret}.

3.1 Modeling the Discharging Loss

For the sake of a pointed explanation, we focus on the discharging loss function f^{ret} and assume the functions f^{self} and f^{sto} for self-discharging and charging, respectively, to be constant. (In practice, f^{self} can safely be assumed to be constant as long as the ambient temperature stays constant. Formulas for f^{sto} are, however, difficult to obtain without detailed information on the situation. Depending on several factors, such as temperature, battery age, current load level, charging regimes can be devised that result in a small charging loss factor. See, for instance, [9–11].) For the discharging loss we consider a strictly monotonically increasing function $f^{\text{ret}} : \mathbb{R}_{\geq 0} \to \mathbb{R}_{\geq 1}$ only depending on the amount of retrieved energy e_t^{ret}. Under these conditions, the discharging factor can be expressed analytically as follows (see, for instance, [12, 13]):

$$f^{\text{ret}}(e_t^{\text{ret}}) = 1 + \frac{c\left(1 - e^{-kg(e_t^{\text{ret}})}\right)}{kg(e_t^{\text{ret}})}, \tag{13}$$

where the auxiliary function g cannot be described elementarily and is implicitly defined by

$$e_t^{\text{ret}} = \frac{Qk}{4c\left(1 - e^{-kg(e_t^{\text{ret}})}\right) + 4kg(e_t^{\text{ret}})}. \tag{14}$$

Q is the maximum capacity (at infinitesimal current), k is a constant rate, and c is the gap between available charge capacity and total capacity. The resulting curve of the function f^{ret} is illustrated in Fig. 2 (solid black line).

Exact Loss Function

Including the exact nonlinear function directly yields a model which is algorithmically hard to solve. Heuristics can be used to improve solvability. To find start solutions, the variables e_t^{ret} can be fixed to either 0 or values depending on the energy costs or the former peak transmission, and the resulting easier (linear) problem can be solved. For improving solutions, a local neighborhood search can be applied to the values of e_t^{ret} of the current best solution.

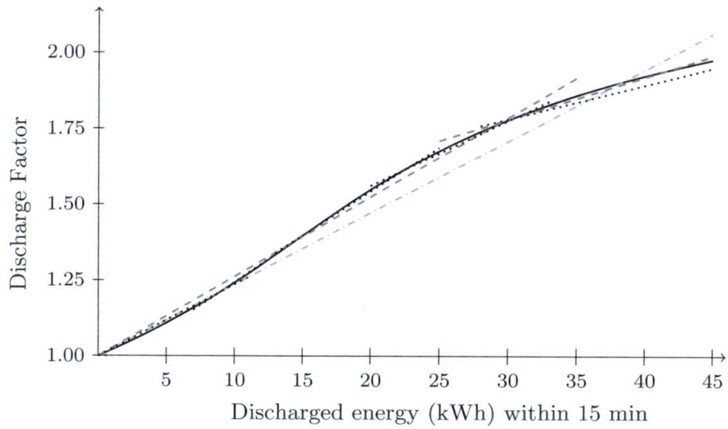

Fig. 2 Discharge factor f^{ret} with respect to the discharging rate at constant potential (*solid*) and linearizations by an affine (*dashdotted*), a concave, piecewise linear function (*dashed*), and a nonconvex, nonconcave, piecewise linear function (*dotted*)

Constant Loss Factor

The roughest way to avoid the nonlinearity is to approximate the function with a constant. Then, we have the same loss factor for each time step, independent of the amount of discharged energy. A special case of this approximation is to use $f^{\text{ret}} \equiv 1$, this is, we assume no losses when discharging the battery.

This constant approximation is the most commonly used discharging loss in the literature, see [3–5].

Iterative Solution with Constant Loss

It is possible to avoid the nonlinearity in the first place by using a constant loss f_t^{ret} for each time step t. In this case the values f_t^{ret} used for the optimization potentially deviate strongly from the real loss. The optimization model is solved repeatedly until a given maximal number of iterations is reached. After each iteration the new solution is used to update the constants f_t^{ret} according to the correct loss. Additionally, the variables e_t^{ret} are bounded depending on their value in the new solution and the former bounds.

Affine and Piecewise Linear Approximation

A finer way of approximating f^{ret} is to use an affine, continuous, concave, piecewise linear, or a continuous, nonconvex, nonconcave, piecewise linear function. All of these variants result in a quadratic or polynomial model, in which the difference between the computed and the real loss corresponds to the gap between the nonlinear discharging factor and the approximation.

These approximations of the nonlinear functions are illustrated in Fig. 2. The dashdotted line is the affine linear approximation, the dashed is the affine, continuous, concave, piecewise linear one, and the dotted line is the continuous, nonconvex, nonconcave, piecewise linear approximation.

Table 4 Parameter values used for the computations

Parameter	Value
$t_{\max}$	96
$s^{\max}$	360 kWh
$s^{\min}$	180 kWh
s^{init}	252 kWh
u^{buy}	∞
u^{sell}	∞
u^{ret}	45 kWh
u^{sto}	9 kWh
f^{sto}	0.9
f^{self}	0.9995
Q	436.51 kWh
k	0.6750
c	1.4363
d, p	see Fig. 3

Since the approximated versions of the discharge functions do not model the discharge exactly, it is possible that the obtained schedules slightly violate the exact side constraints. Therefore, in a postprocessing step violations, such as a load level of more than 100 % or less than the allowed minimal level, are repaired and correctness with respect to the exact constraints is ensured.

The usage of the exact discharge loss function results in a mixed integer nonlinear program (MINLP) that has to be solved, whereas the various approximations lead to mixed integer quadratic programs (MIQP)—or can at least be reformulated as such—or even to mixed integer linear programs (MILP). These models can be solved with a suitable optimization software, such as SCIP [14, 15].[2] Usually, the solution process involves solving a linear relaxation and then continuing in a branch-and-bound phase, where a branch-and-bound tree is implicitly generated whose nodes correspond to suitable subproblems. For more details on branch-and-bound see, for instance, [16, 17].

To speed up the computation, an *early termination strategy* can be used, where the solving process is stopped after 1000 branch-and-bound nodes of stalling, i. e., nodes without an improvement upon the best obtained solution so far. This solution is possibly not optimal, but in general it can be expected that continuing the solving process does not lead to better solutions in a reasonably short time.

In the following, we compare these different approaches with respect to solvability (run-time) and quality of the solution. The parameter values used for the computations are given in Table 4 and Fig. 3 and are suited to model the scenario of a small ICT central office location.

[2] see http://scip.zib.de.

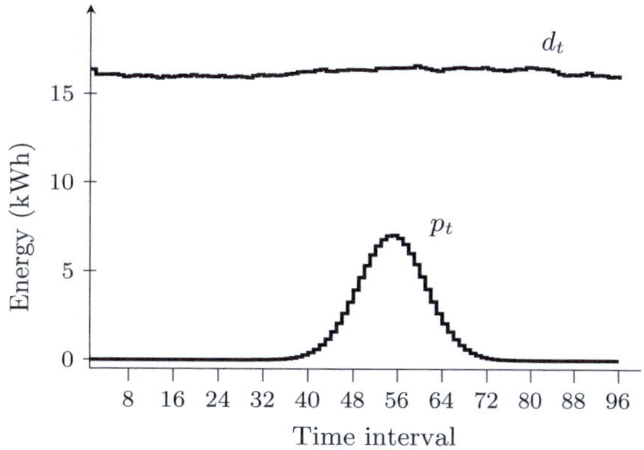

Fig. 3 Parameter values for produced energy from solar panels (p_t) and demanded energy (d_t) in each time step for an exemplary sunny day

3.2 Case Study: Optimizing Energy and Transmission Costs

We consider an example where we minimize both energy costs and peak increase. The target peak transmission was set to 80 kW. Energy prices are taken from the spot market of the EEX (European Energy Exchange) in Leipzig.[3] Net demand—consumed energy at the node minus the produced amount of energy—decreases during the day, due to electricity produced by solar panels during daylight. Figure 4 illustrates a schedule for charging/discharging and trading energy on an exemplary day.

3.3 Computational Study

In this section we report on computational studies on 60 scenarios. Each scenario represents a whole day (from 0:00 to 24:00). The consumer's energy demand is assumed to be composed of the consumption of two types of electrical devices:

- permanently powered devices—constantly running hardware such as ICT equipment at a central office or fridges at a household
- daylight devices—such as lights, personal computers or general household appliances that are usually switched off during the night

The 60 scenarios consist of three groups, each containing 20 scenarios, of the following types:

- producers with hardware of both of the above types and a varying energy-production over day (standard smart grid nodes)
- non-producers with both types of devices, but without energy-production (standard power-consuming locations)

---

[3] see http://www.eex.com/en#/en.

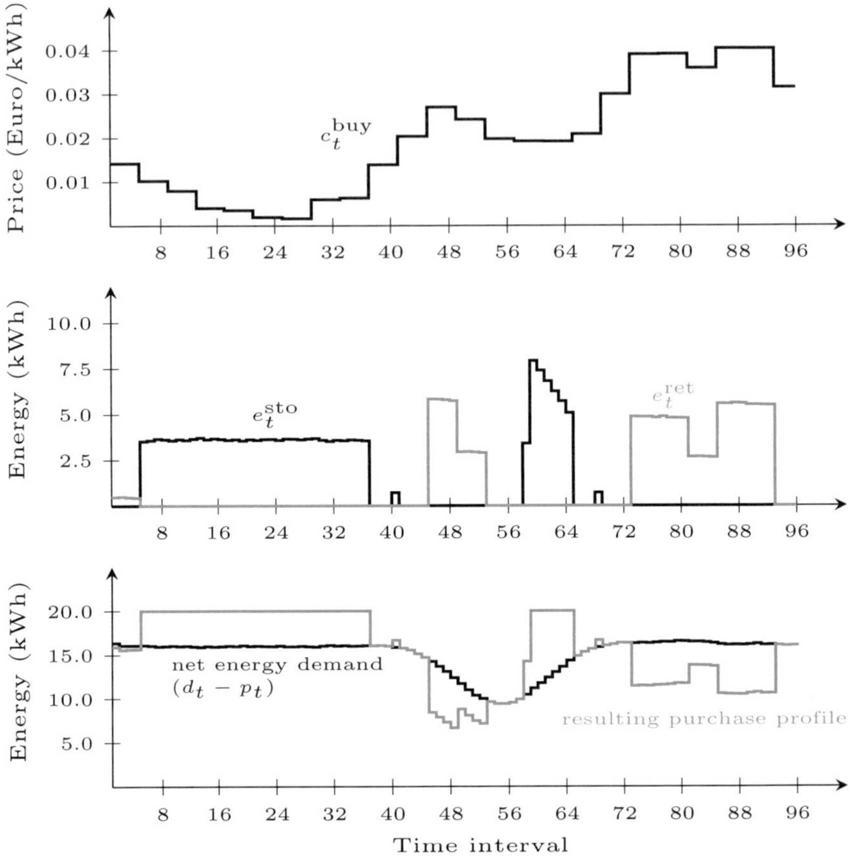

Fig. 4 Optimal schedule for an example day: Energy prices (*top*), amount of energy stored in/retrieved from the battery (*middle*), and net energy demand and resulting energy purchase profile (*bottom*). Energy is stored in the battery (*dark curve, middle*) during the night and over 2 h in the afternoon, when prices are low, and retrieved to power the site (*light curve, middle*), when prices are higher. At the same time, the increase over the preset peak value of 80 kW (corresponding to 20 kWh of energy in every quarter hour) is aimed to be kept small—indeed, the resulting purchase profile (*light curve, bottom*) shows that the peak is not increased at all, i.e., too high peaks in the purchase profile are shaved off. The impact of the physical properties of the battery and their modeling can also be read off from the figure. Due to the energy losses involved, the integral of the *middle dark curve* (the total stored energy) is smaller than the integral of the *middle light curve* (total retrieved energy). The more precise the used technical constraints model the losses, the more reliable is the resulting schedule. Furthermore, a higher discharge curve yields an even greater difference of these integrals, due to the increasing discharge loss function; hence high discharge peaks are avoided: in this example the discharged energy in every quarter hour is always significantly below the maximally allowed value of 45 kWh (cf. Table 4)

- non-producers with flexible power consumption for the permanently powered devices (the role model for such locations being future central office sites with load-adaptive ICT equipment)

The power consumptions of the devices may depend on the type of day (working or weekend day), the outside temperature, and the production depends on the sunshine duration of the day. Twenty randomly chosen days throughout the course of the year 2011 are considered, each with historic temperature data and energy prices from the European electricity spot market EEX.

All computations are carried out with a time limit of 300 s, using the solver SCIP [14] on a 64 bit Linux system running on an 8-core Intel Xeon X5672 3.2 GHz CPU with 12 MiB cache and 48 GiB RAM (each run uses only one core).

Table 5 lists numerical results. In the columns of the table, we list solutions with respect to the four objectives presented in Sect. 2.2. In the blocks (A)–(G) of the table rows, we list results for seven approaches discussed in Sect. 3.1. Each block shows the objective values of the repaired solutions as well as the running times of the respective methods, where every number is the shifted geometric mean[4] over the results from all 60 scenarios. We list cost function values and running times for the regular optimization process with time limit as well as for our early termination heuristic (see Sect. 3.1). For comparison, the topmost block (0) shows the mean values for the 0-schedules, i. e., when no battery activity is taking place.

Comparison of the Approaches

The approximation variants (C) and (D) come out as the overall favourable solution approaches with—at least for the first three objectives—the iterative method (E) as strong alternative. These three approaches yield comparably good solutions on average. The iterative method (E) is fastest in general, while the other two methods suffer from some time-outs. Using the exact loss function (A) or approximations with many affine functions (B) leads to worse solutions within the given time limit. These approaches are more precise, but result in models that are much harder to solve. Longer run-times would hence be necessary for obtaining better or even optimal solutions. On the other hand, the constant approximation variants (F) and (G) can be solved quickly, but often result in solutions of mediocre quality. Nevertheless, compared to the 0-solution (0), all versions provide improvements for nearly all objective functions, i. e., optimization is paying off.

Influence of the early termination strategy

In order to reduce running times, the early termination strategy can be employed. It stops the solution process after 1000 branch-and-bound nodes of stalling. For approaches (C) and (D), this seems to be very effective and reduces the run-times and number of time-outs considerably. The early termination strategy has no influence on the iterative approach (E), since the MIPs solved in the iterations are relatively easy and do not produce sufficiently large branch-and-bound trees. The same happens

---

[4]The shifted geometric mean of n values $a_1, \ldots, a_n \geq 0$ is defined as

$$\left(\prod_{i=1}^{n} (a_i + s)\right)^{\frac{1}{n}} - s,$$

where $s > 0$ is a sufficiently small shift parameter.

Table 5 Computational results for the seven different solution methods with four different objectives

Objective	Energy costs (€)	Peak power increase (kW)	Total costs (€)	Target profile (total deviation)
(0) Without optimization				
Solution	76.831	8.342	100.157	317.037
(A) Exact function for discharge factor				
Best solution	75.800	2.934	85.579	192.158
Time (out)	300.000 (60)	2.781 (4)	283.374 (59)	300.000 (60)
Early termination	75.800	2.934	85.579	192.538
Time (out)	11.631 (0)	0.552 (0)	12.988 (0)	14.632 (0)
(B) Piecewise linear approximation of discharge factor by 5 affine functions				
Best solution	76.054	2.765	84.390	177.786
Time (out)	300.000 (60)	2.725 (2)	300.000 (60)	300.000 (60)
Early termination	76.056	2.934	85.784	182.362
Time (out)	38.123 (0)	1.456 (0)	34.798 (0)	47.362 (0)
(C) Piecewise linear approximation of discharge factor by 2 affine functions				
Best solution	75.008	2.608	83.076	167.396
Time (out)	20.722 (7)	0.741 (1)	16.420 (5)	108.351 (38)
Early termination	75.008	2.608	83.076	172.083
Time (out)	4.274 (0)	0.187 (0)	5.577 (0)	5.542 (0)
(D) Linear approximation of discharge factor				
Best solution	75.005	2.605	83.059	168.080
Time (out)	25.128 (10)	0.743 (1)	19.860 (7)	120.954 (42)
Early termination	75.005	2.605	83.060	173.489
Time (out)	4.200 (0)	0.201 (0)	5.839 (0)	5.724 (0)

(continued)

Table 5 (continued)

Objective	Energy costs (€)	Peak power increase (kW)	Total costs (€)	Target profile (total deviation)
(E) Iterative solution with constant discharge factor				
Best solution	75.092	2.631	83.136	190.337
Time (out)	0.715 (0)	0.419 (0)	0.761 (0)	58.242 (0)
Early termination	75.092	2.631	83.136	190.337
Time (out)	0.748 (0)	0.437 (0)	0.799 (0)	58.239 (0)
(F) Constant approximation of discharge factor				
Best solution	76.374	2.631	84.346	175.798
Time (out)	0.126 (0)	0.109 (0)	0.234 (0)	41.054 (20)
Early termination	76.374	2.631	84.346	175.763
Time (out)	0.128 (0)	0.101 (0)	0.249 (0)	3.700 (0)
(G) Zero discharge factor				
Best solution	77.008	3.175	85.321	176.826
Time (out)	0.138 (0)	0.105 (0)	0.291 (0)	32.961 (12)
Early termination	77.008	3.175	85.321	176.823
Time (out)	0.138 (0)	0.098 (0)	0.304 (0)	4.895 (0)

Shifted geometric means over 60 instances of objective values of best obtained solutions and best solutions obtained after early termination, as well as run-times for the respective solution processes (and number of runs in each section that hit the time limit of 300 s) for the constant approximations (F) and (G), except when aiming at a given target profile. In the latter case, we have to deal with a nonlinear objective containing absolute values; its linearization blows up the model to some extend and the solution process generates bigger branch-and-bound trees, for which early termination proves more effective.

In summary, early termination significantly reduces the running times to at most a few seconds without causing too much deterioration in solution quality in all important cases. The solutions are computed fast enough so that the model can be used in a rolling-horizon implementation, where it has to be resolved periodically after suitable time intervals, such as 15 min.

Effect of repairing solutions

We finally point out a somewhat counterintuitive effect that results from the post-processing necessary for the approximation variants. Here, the final (repaired) solutions after early termination might have a slightly better objective value than the best solutions found in the full optimization run. In Table 5 this effect even shows in the mean value for the constant approximation approach (F) when minimizing the deviation from a target profile.

4 Conclusion and Outlook

We have presented a flexible approach based on mathematical optimization for computing optimal schedules for trading and buffering energy in a smart grid. We have applied the approach to a smart grid scenario, using batteries as storage devices. The techniques have been successfully implemented into a control system for a smart grid node, optimizing storage schedules every 15 min.

We studied alternative objective functions, such as minimization of costs or deviation from a target load profile. By varying the level of detail in modeling technical constraints, we observe a trade-off between solvability of the model and the quality of solutions.

The presented model can be extended in various ways. Additional constraints can be incorporated, accounting for maintenance of the storage device and special contracts with power suppliers. Future extensions may address the involvement in the trade of operating reserve electricity and other storage technologies, such as pumped-storage hydroelectricity.

Acknowledgments This work was supported by the German Federal Ministry of Economics and Energy (BMWi) within the DESI project (led by the Telekom Innovation Laboratories) as part of the IT2Green framework (`it2green.de`). The authors thank Dr. Heiko Lehmann (Telekom Innovation Laboratories) for his support and guidance as well as the reviewers for constructive comments and suggestions.

References

1. Dunn, B., Kamath, H., Tarascon, J.M.: Electrical energy storage for the grid: a battery of choices. Science **334**(6058), 928–935 (2011)
2. Kondoh, J., Ishii, I., Yamaguchi, H., Murata, A., Otani, K., Sakuta, K., Higuchi, N., Sekine, S., Kamimoto, M.: Electrical energy storage systems for energy networks. Energy Convers. Manag. **41**, 1863–1874 (2000)
3. Exarchakos, L., Leach, M., Exarchakos, G.: Modelling electricity storage systems management under the influence of demand-side management programmes. Int. J. Energy Res. **33**, 62–76 (2009)
4. Sousa, T., Morais, H., Vale, Z., Faria, P., Soares, J.: Intelligent energy resource management considering vehicle-to-grid: A simulated annealing approach. IEEE Trans. Smart Grid 3(1) (2012)
5. Stluka, P., Godbole, D., Samad, T.: Energy management for buildings and microgrids. In: 50th IEEE Conference on Decision and Control and European Control Conference (CDC-ECC) (2011)
6. Barbato, A., Carpentieri, G.: Model and algorithms for the real time management of residential electricity demand. In: Energy Conference and Exhibition (ENERGYCON), pp. 701-706 (2012)
7. Fischetti, M., Sartor, G., Zanette, A.: A MIP-and-refine matheuristic for smart grid energy management. Int. Trans. Oper. Res. **22**(1), 49–59 (2015)
8. Zhang, D., Papageorgiou, L.G., Samsatli, N.J., Shah, N.: Optimal scheduling of smart homes energy consumption with microgrid. In: 1st International Conference on Smart Grids, Green Communications and IT Energy-aware Technologies (2011)
9. Berndt, D.: Temperaturabhängigkeit der Stromaufnahme beim Laden von Pb- bzw. PbO_2-Akkumulatoren. Chem. Ing. Tech. **38**(6), 627–630 (1966)
10. Dambrowski, J.: Lade- und Diagnoseverfahren für Bleibatterien in unterbrechungsfreien Stromversorgungen–Teil 1. Elektronik 21 (2009). http://www.elektroniknet.de/power/energiespeicher/artikel/1569
11. Dambrowski, J.: Lade- und Diagnoseverfahren für Bleibatterien in unterbrechungsfreien Stromversorgungen–Teil 2. Elektronik 26 (2009). http://www.elektroniknet.de/power/energiespeicher/artikel/1570
12. Jongerden, M.R., Haverkort, B.R.: Which battery model to use? IET Softw. **3**(6), 445–457 (2009)
13. Manwell, J.F., McGowan, J.G., Abdulwahid, U., Wu, K.: Improvements to the hybrid2 battery model. In: Windpower 2005 Conference. American Wind Energy Association (2005)
14. Achterberg, T.: SCIP: solving constraint integer programs. Math. Program. Comput. 1(1), 1-41 (2009). doi:10.1007/s12532-008-0001-1
15. Vigerske, S.: Decomposition of multistage stochastic programs and a constraint integer programming approach to mixed-integer nonlinear programming. Ph.D. thesis, Humboldt-University Berlin (2013)
16. Achterberg, T., Koch, T., Martin, A.: Branching rules revisited. Oper. Res. Lett. **33**(1), 42–54 (2005)
17. Dakin, R.J.: A tree-search algorithm for mixed integer programming problems. Comput. J. **8**(3), 250–255 (1965)

Pre-operative Activities and Operating Theater Planning in Emilia-Romagna, Italy

Andrea Lodi and Paolo Tubertini

Abstract The high costs of health care push national and regional health services as well as local authorities to constantly improve management performances in order to obtain more efficient organization of hospitals activities and provide patients with the best possible care. During the last decade the number of research studies that aim at efficiently organizing and planning surgical activities by reducing costs and keeping a good level of care has dramatically grown. We present a case-study focused on a local hospital department that has to jointly plan operating room assignment to surgeons and patients admission by taking into consideration, for a subset of patients, some compulsory pre-operative activities such as outpatient anesthetist appointment and autologous blood donation. We propose a Mixed Integer Programming model, we test it on real-world instances and we compare the results with the currently used planning procedures.

Keywords Resource scheduling · Integer programming · Healthcare management

1 Introduction: Operating Theater Planning

The high costs of health care push national and regional health services as well as local authorities to constantly improve management performances in order to obtain more efficient organization of hospitals activities so as to provide patients with the best possible care. Operating Theater (OT) planning is one of the most studied topics for Operations Research applied to health systems management. Surgery departments play a crucial role in hospital organization and surgical activities usually constitute the core of efforts made by hospital managers in order to improve performances. Surgery activities represent a significant cost in the overall hospital budget because they directly involve expensive resources but also because they have an impact on other

A. Lodi (✉) · P. Tubertini
DEI—Università di Bologna, Viale Risorgimento 2, Bologna, Italy
e-mail: andrea.lodi@unibo.it

P. Tubertini
e-mail: paolo.tubertini@unibo.it

K. Fujisawa et al. (eds.), *Optimization in the Real World*,
Mathematics for Industry 13, DOI 10.1007/978-4-431-55420-2_7

hospital services and on a multitude of resources indirectly associated. During the last decade the number of research studies that aim at efficiently organizing and planning surgical activities by reducing costs and keeping a good level of care has dramatically grown. A better planning and management of activities and resources directly and indirectly involved in surgical procedures (Operating Theater Planning) can therefore entail a more efficient use of resources, a reduction in waiting times for patients and a better overall performance of the hospital itself. In this article we consider the problem of assigning Operating Room (OR) time slots to surgeons in parallel with the definition of the admission planning of elective patients. The proposed approach then combines tactical and operational decisions aiming at optimizing the resources utilization in relation with surgical waiting lists. The paper is organized as follows. In Sect. 2 we review the literature with the various elements that have been considered in OT optimization models. In Sect. 3 we propose a Mixed Integer Programming (MIP) formulation based on Emilia-Romagna guidelines and customized on a local hospital orthopedic department. Section 4 reports the computational results for real-wold instances. Finally, Sect. 5 draws some conclusions and outlines some future work.

2 Literature Review

Articles on Operating Theater management are characterized by a wide range of different criteria and approaches. Extensive reviews on OT planning models can be found in [1, 2]. Both articles report a vast number of studies in which OT planning has been analyzed and solved by means of Operations Research techniques. Several Mixed Integer Programming formulations have been proposed and solved with exact or heuristic procedures. Nevertheless a consistent branch of research applies simulation techniques in order to evaluate the dynamic behavior of the system under different configurations. The goal of the proposed literature review is first to resume the characteristics of the three decision levels, strategic, tactical and operational, in terms of planning horizon and decision-maker general goals. Then, to identify the characteristics of patients and resources so as to define the boundaries of the planning problem. Finally, to compare planning objectives in terms of patient care and organizational efficiency.

2.1 Strategic Planning

Strategic planning decisions usually aim at defining the number of operating rooms that will characterize future hospital structures. Decisions concerning the amount of investments for the construction of new operating rooms as well as for technical equipment procurement and the prediction of specialists' need are characteristic elements of strategic long-term planning. Furthermore, welfare policies defined by local or regional health management are discriminating factors for capacity planning activities as well as patient mix management. Examples of strategic planning can be found in [3, 4].

2.2 Tactical Planning

Strategic decisions, despite having a strong impact on operational performance, cannot be considered as the only decision that can be managed with quantitative tools. A proper management of available resources can, in some cases, avoid investments for facility upsizing. Tactical OT planning, also known as Master Surgical Schedule Problem (MSSP), focuses on the assignment of available scarce resources to surgery teams. The Master Surgical Schedule determines the assignment of OT time slots to surgery teams on a weekly basis and it is usually considered valid for a 1 year time horizon, i.e. it is cyclically applied rather than modified week by week. The MSSP is designed to support the negotiation phase among medical specialization units involved in order to ensure the most equitable resource distribution. Examples of tactical planning can be found in [5–7].

2.3 Operational Planning

Operating Theater operational planning has to identify, for each medical specialization unit, which patients, among the ones in the waiting list, will undergo surgery during the week. This decision-making phase is characterized by short-term (weekly or daily) goals and it is usually called Admission Planning Problem (APP) or Surgical Case Assignment Problem (SCAP) if it includes tactical decisions regarding OR time slot assignment to surgeons (see [8]). Examples of operational planning model can be found in [6, 9–11]. Operational planning can be influenced by various objectives. As an example, in [12] the authors define patients categories and determine how many patients, of each category, have to undergo surgery in order to obtain the best patient's acceptance profile. A different approach can be found in [13] where the authors take into account patients priority according to their clinical status and the time spent on the waiting list. Finally, human and material resource utilization during surgery is analyzed in [14] while the impact of operating room planning on pre- and post-surgery resource utilization is evaluated in [15].

2.4 Patients Characteristics

A first distinction of OT problems can be driven by the set of patients that are considered during the planning phase. In the literature, two classes of patients are mentioned: elective patients and non-elective ones. We define an elective patient as a person who has to undergo a surgery that can be planned in advance (elective surgery). We define non-elective patients as those who need to urgently undergo surgery due to unexpected events that cannot therefore be planned in advance. It is possible to split elective patients in two categories: inpatients, namely people that need to be hospitalized for at least one night, and outpatients that enter and leave the hospital on the same day of the surgery. In [12] the authors develop a MIP model for scheduling elective operations, where outpatients are treated as inpatients requiring a hospital stay of only 1 day, the one in which the surgery takes place. Among non-elective surgeries, however, it is possible to distinguish urgent and emergency ones depending on the timeline with which they have to be performed which, in turn, depends on the patient clinical condition. Thus, elective surgeries can be planned in advance without any kind of uncertainty related to the time of arrival, while non-elective ones are hardly predictable due to the stochastic behavior of urgencies and emergencies (see [16, 17]). Most of the research dealing only with OT planning considers mathematical models in which a set of ORs are dedicated to non-elective surgeries in order to avoid disruptions of elective activities. Alternative approaches to non-elective activities management have been proposed in order to reduce OR underutilization. In [18] the authors studied, by means of Discrete Event Simulation (DES) technique, the impact of reserving daily time slots to non-elective surgeries and concluded that the overall use of operating rooms improves significantly when the OT capacity dedicated to non-elective surgeries is distributed across operating rooms.

2.5 Resource Constraints

Among the resources directly involved in surgical procedures operating rooms are, of course, the most important ones. In [5] the authors compare two methods of managing ORs time capacity. In the first model they manage OR time as a fixed resource constraint in which, on each day, the sum of surgery hours has to be less than or equal to the regular time capacity. In the second model, instead, the proposed schedules might include some overtime with respect to classical operating room capacity. Bounds on regular time and overtime OR capacity are defined in order to forbid an overload of activities. Similar constraints can be found in the three stage approach of [7]. Surgery team availability is taken in consideration by [5, 11] requiring that, in each day, surgeries are carried out only if there is at least one surgeon available. In [11], as well as in [7, 13, 19] the authors define non-ubiquity constraints related to surgeons or to surgical staffing units, where a non-ubiquity constraint is the maximum number of simultaneous activities that a resource can do.

So far only resources directly related to surgery activities have been described. It is clear that there is a set of hospital resources that are indirectly influenced by Operating Theater planning. Ward and Intensive Care Unit (ICU) beds usually define hard constraints to deal with since surgeries cannot be scheduled if no bed is available for post-operative stay. In [19] the authors restrict the number of patients that can undergo surgery in a day, depending on the number of beds available in ward and ICU. In addition to capacity constraints some models consider also political or legal constraints. In [5] the authors impose lower and upper bounds on the number of time slots that can be assigned to each specialization. In [6] the authors set a limit to the maximum number of hours that a surgeon can work in a day.

A further element of distinction among articles dealing with OT planning is the variability associated with surgery duration. In [20] the authors propose a stochastic integer programming model with a set of probabilistic constraints requiring that the total duration of all procedures performed in a day, in a certain operating room, cannot exceed OT capacity, with a given probability. A robust model is presented in [21] to manage the stochastic behavior of surgery durations.

2.6 Objective Function: Performance Indicators

In order to define if a given schedule or assignment is more efficient than others, it is necessary to identify performance indicators, and several have been considered in the literature. It is possible to identify two sets of objectives that are taken into account.

The first is characterized by those goals that aim at improving the service provided to patients by reducing waiting times and by taking in consideration patients needs and priority related to their clinical status. Long waiting lists are one of the most common reported problem in health services. This results in a large quantity of studies that aim at reducing the waiting time of patients in order to increase their degree of satisfaction. This kind of objective can be identified in [11, 19], where the authors evaluate the overall efficiency of planning as a function of patients waiting time.

The second is focused on the improvement of hospital organizational efficiency in terms of costs minimization and resources utilization maximization. In this context, [9] proposes a model that aims at maximizing the number of patients who may be hospitalized. In [20] the authors try to increase OT productivity by reducing inactivity times between two consecutive surgeries. Resource utilization is one of the most used performance indicators that focus on hospital organizational efficiency. In [15] the authors develop a MIP model that minimizes the deviation between the target utilization of resources (e.g., beds, nursing staff, and number of surgeries that should be performed during the planning horizon.) and their actual use. The decision to perform surgeries in overtime may be more convenient in terms of waiting time reduction. On the other hand, keeping an operating room open can be very expensive and, for that reason, the number of planned surgeries beyond the regular schedule should be kept under control (see [5, 11]). Resource leveling is another important goal for OT management. For example, the prevention of peaks in resource utilization is taken in consideration in [20].

3 Operating Theater Management for a Local Hospital Orthopedic Department

Emilia-Romagna Region lacks of optimization-driven decision support systems for OT management. Most of local hospitals are equipped with data management tools that, at the best of our knowledge, are mainly used to record activities rather than to plan them. Furthermore, local databases sometimes lack of fundamental planning information such as surgery duration, because the adoption of data management systems is in some cases not enforced.

Implementing an OT model requires the collection of a large set of data in order to feed model parameters and properly defining its coefficients. Tests can be performed and analyzed using randomly generated or real-world data. Once the model based on the regional guidelines has been designed, we had the opportunity of testing it on real-world data instances. An orthopedic department of a local health hospital agreed to collaborate with us in terms of model validation and data supply. This is of course a much more favorable situation because the feedback from health care experts is priceless.

The proposed model is focused on department internal dynamics and considers assignment of operating rooms among surgical units as a strategic decision that is taken by hospital managers and that cannot be modified by department managers. The problem we analyze is then defined as the allocation of OT time slots, among the ones already assigned to the department, to orthopedic surgeons and the subsequent admission plan for their related patients.

All surgeons working at the department share a common waiting list of patients that has been previously examined within the public health service. In addition, each surgeon has a personal list of patients waiting for surgery. The final problem can then be defined as the determination of which OT time slots will be assigned to each surgeon and which patients, belonging to shared or personal lists, must undergo surgery in the planning week. Orthopedic surgeries can be classified in elective and non-elective ones. Elective surgeries can be distinguished in non-prosthetic and prosthetic. Prosthetic surgeries consist of articulations replacement due to disease or congenital conditions and can be classified in hip replacement, knee replacement, and shoulder replacement. Following the input of physicians we decided to overlook non-elective surgeries, essentially trauma surgeries, since trauma management is organized by reserving on a daily basis OT time slots.

In summary, with respect to the literature review performed in Sect. 2, the keywords of our case-study are

- Hybrid tactical and operational OT planning (SCAP);
- Pre-operative outpatient consultations.

3.1 Problem Description

The Orthopedic department has a total of 20 reserved time slots per week, 4 of which are reserved to trauma management. Each day is partitioned into two periods: a morning session of 6 h from 8 AM to 2 PM and an afternoon one of the same duration, from 2 PM to 8 PM. A feasible schedule dedicates to each specific surgeon one or more day periods during the planning horizon. An operating room, therefore, is occupied by a single surgeon for the duration of the assigned period. In addition, the Orthopedic department managers impose that, in each period, only one class of surgeries can be performed (prosthetic, non-prosthetic or trauma). Finally, it is preferable that prosthetic surgeries are performed during the morning session. As already mentioned, the prosthetic interventions are classified as hip, knee or shoulder surgery. Each type is characterized by a distinct degree of complexity, so it is preferable, for organizational reasons, to minimize the number of different types of prosthetic surgeries that are performed in the same time slot. As a standard description we can consider that two surgeries per time slot can be planned. However, it is possible to plan on overtime an additional surgery just for non-prosthetic ones. (Note that this assumption of two surgeries per time slot applies to our case study and not to a generic OT model.) Finally, it is possible that, for specific surgeries, a surgeon should be supported by a colleague. The model then must satisfy these requests by preventing the assignment of an operating room to a surgeon, in case he has to assist a colleague.

Once activities that directly involve Operating Theater have been analyzed we evaluated which could be the set of resources that could indirectly impact OT planning. Beds availability has not been considered in the model since the number of beds currently dedicated to the orthopedic department appears to be relatively abundant if compared to historical demand and therefore it does not constraint surgical activities.

Pre-operative management of outpatient anesthetic examinations is instead an element of great importance while planning surgical activities. Patients waiting for surgery are characterized by an heterogeneous level of urgency, associated with their clinical condition. Each patient can then be classified with a degree of potential criticality that, in cases of major significance, can require a pre-operative anesthetic consult in addition to the routine one. This medical examination is associated with the possible, but not certain, need of in-depth medical examinations in order to give to the anesthesiologist a complete clinical picture of the patient, so as to properly manage his peri-operative clinical pathway. Inefficiency may arise when a surgical procedure cannot be performed because some of the data necessary in order to preserve patient's health condition are missing. This situation can lead to surgery cancellations that negatively impact both patient care and hospital resource utilization. Usually, anesthesiologists define, for each medical specialization unit, the maximum number of pre-operative outpatient medical examinations that can be planned. In case it has not been possible to book an anesthetist examination for a critical patient, hospitalization is increased from 1 to 4 days in order to ease an informal and more direct management of anesthetist evaluation. In a similar way autologous blood donation is a pre-operative activity that must be planned in relation with surgical activities. Autologous blood donation is strongly promoted by national and regional guidelines. Indeed, since the planning of the blood drawing samples of patients who may have to undergo blood transfusions during surgery is important for several reasons. Autologous blood donation can lead to patient self-sufficiency in terms of blood needs during a surgery and can also reduce the volume of requests to hospital blood bank that has a limited number of blood bags shared by all the medical specialization units in the hospital. Autologous blood donation is characterized by two blood samples, the first that has to be organized at least 3 weeks before the surgery day and the second that must be taken at least 1 week before it.

3.2 Planning Objectives

Once defined the optimization problem and the set of constraints that will influence the definition of a feasible solution, it is important to identify which are the indicators to be used in order to evaluate solutions and to identify the most appropriate one. The concept of appropriateness in health care is not always uniquely defined and quantitatively measurable. We decided to take into account both types of objectives that have been reviewed in Sect. 2, namely, patient-related quality of service and hospital-driven internal efficiency.

Patient-driven indicators

Quality of care and patient centered treatment are considered fundamental driving factors for the Emilia-Romagna public health care system and time-to-surgery can be considered as the most important one. We define time-to-surgery as the lead time between patient referral and the day of surgery, where patient referral is defined as the moment in which, after diagnosis, the surgeon decides that the patient needs a surgical treatment. Time-to-surgery is then an indicator of health system responsiveness to population needs and is strongly influenced by patient clinical status. Patient clinical condition, pathology evolution risk and surgery complexity are used at a regional level to define priority classes for elective surgeries: (a) High priority: Time to surgery has to be bounded by 30 days; (b) Medium priority: Time to surgery has to be bounded by 90 days; (c) Low priority: Time to surgery has to be bounded by 360 days, and (d) Very low priority: Time to surgery has no bound and it can exceed 360 days.

Hospital driven indicators

Hospital-driven indicators are related to an efficient resource management. We considered two elements of organizational evaluation, namely, the number of overtime hours (to be minimized as much as possible) and the number of pre-operative activities (to be maximized as much as possible).

3.3 Problem Formulation

Let I be the set of patients inserted in the waiting lists, K the set of ORs, W the set of surgeons and T the set of days in the planning time horizon. We consider as planing time horizon a period T equal to 5 weeks and we define the last five working days of period T as the ones where surgery activities will be planned. We define working days as the ones available for assignment because planned surgeries are usually performed from Monday to Friday. Each working day is subdivided in P time slots (namely, $P = 2$, usually morning and afternoon session). Let us define Q as the set of surgery cohorts, namely prosthetic and non-prosthetic surgeries and R_q the subset of surgery types that characterize $q \in Q$ class of surgeries. The model we propose considers as a first decision variable the assignment of OT time slots to surgeons as

$$g_{wktpq} = \begin{cases} 1 \text{ if time slot } p \in P \text{ on day } t \in T \text{ of operating room } k \in K \text{ is assigned} \\ \quad \text{to surgeon } w \in W \text{ for surgeries of class } q \in Q \\ 0 \text{ otherwise.} \end{cases}$$

To distinguish if more than one type of surgery is performed we define variables

$$y_{wktpr} = \begin{cases} 1 \text{ if during time slot } p \in P \text{on day } t \in T \text{ of operating room } k \in K \\ \quad \text{a surgery of type } r \in R_q \text{ will be performed by surgeon } w \in W \\ 0 \text{ otherwise.} \end{cases}$$

Each specialization time slot assigned is then filled with patients that are related to the surgeon waiting list as

$$x_{iktp} = \begin{cases} 1 \text{ if patient } i \in I \text{ is assigned to operating room } k \in K \text{ on day } t \in T \\ \quad \text{in time slot } p \in P \\ 0 \text{ otherwise.} \end{cases}$$

Since not all patients in the waiting list can undergo surgery during the week of planning we define

$$z_i = \begin{cases} 1 \text{ if patient } i \in I \text{ will not undergo surgery on time horizon T} \\ 0 \text{ otherwise.} \end{cases}$$

Pre-surgery anesthetist appointments will be managed by variables

$$\varphi_{it} = \begin{cases} 1 \text{ if patient } i \in I \text{ is examined by an anesthetist on day } t \in T \\ 0 \text{ otherwise;} \end{cases}$$

$$\varepsilon_i = \begin{cases} 1 \text{ if patient } i \in I \text{ will undergo surgery and no anesthetist examination} \\ \quad \text{is planned} \\ 0 \text{ otherwise.} \end{cases}$$

First pre-surgery autologous blood donations will be managed by variables

$$\sigma_{it} = \begin{cases} 1 \text{ if the first pre-surgery autologous blood donation appointment of} \\ \quad \text{patient } i \in I \text{ is planned on day } t \in T \\ 0 \text{ otherwise.} \end{cases}$$

Second pre-surgery autologous blood donations will be managed by variables

$$\lambda_{it} = \begin{cases} 1 \text{ if the second pre-surgery autologous blood donation appointment of} \\ \quad \text{patient } i \in I \text{ is planned on day } t \in T \\ 0 \text{ otherwise;} \end{cases}$$

Table 1 Model parameters

P_{WT}	Objective function waiting time weight
P_{OT}	Objective function overtime weight
P_{SP}	Objective function overlapping assignment
P_V	Objective function unplanned anesthetist appointments weight
P_{AV}	Objective function unused anesthetist appointments weight
P_B	Objective function unplanned autologous blood donation appointments weight
P_{BS}	Objective function unused autologous blood donation appointments weight
I_w	Subset of patients that are in the waiting list of surgeon $w \in W$
I_{ward}	Subset of patients that are in the general waiting list
R_q	Subset of surgery types that characterize class $q \in Q$ of surgeries (e.g., for $q = 1$ three type of surgeries are defined, namely hip, knee and shoulder)
I_r	Subset of patients that need to undergo surgery type $r \in R_q$
Δ_i	Waiting time of patient $i \in I$
$\pi_i^{\Delta_i}$	Priority coefficient of patient $i \in I$ related to deadline proximity (see [22])
ρ_i	Priority coefficient of patient $i \in I$ related to his clinical status
F_{kt}	Maximum number of overtime hours planned for operating room $k \in K$ on day $t \in T$
O_{kt}	Overtime cost of operating room $k \in K$ on day $t \in T$
M	Maximum number of prosthetic surgery types that can be assigned to the same operating room on the same time slot
E_{kt}	Cost related to exceeding number of different medical specialization units assigned to the same time slot for operating room $k \in K$ on day $t \in T$
L_w	Minimum number of surgery hours to be assigned to surgeon $w \in W$
U_w	Maximum number of surgery hours o be assigned to surgeon $w \in W$
L_{ward}	Minimum number of patients coming from the ward waiting list that should be planned in the week
U_{ward}	Maximum number of patients coming from the ward waiting list that should be planned in the week
S_{ktp}	Number of surgical cases that can be treated on day $t \in T$ for operating room $k \in K$ during time slot $p \in P$
a_{ktp}	Equal to 1 if on day $t \in T$ operating room k during time slot $p \in P$ is dedicated to trauma surgeries
μ_i	Clinical condition indicator, equal to 1 if patient $i \in I$ is classified as critical and needs an anesthetist examination before surgery
h_i	Number of days that are suitable between anesthetist examination and surgery time in order to guarantee further in depth examinations if necessary
C_t	Number of anesthetist outpatient appointment slots available on day $t \in T$

(continued)

Table 1 (continued)

ν_i	Clinical condition indicator, equal to 1 if patient $i \in I$ is classified as suitable for autologous blood donation
B_t	Number of autologous blood donation outpatient appointment slots available on day $t \in T$
N	"big M" value used to link x_{iktp} and y_{wktpr} variables
D	Set of supporting surgeries defined as: $d = \langle w_1, w_2, r_d \rangle$ where r_d is the type of surgery for which w_1 needs the support of surgeon w_2
$T1$	Pre-surgery time subset $t \in \{1, \ldots, \lvert T \rvert - 5\}$
$T2$	Surgery time subset $t \in \{\lvert T \rvert - 4, \ldots, \lvert T \rvert\}$

$$\zeta_i = \begin{cases} 1 \text{ if patient } i \in I \text{ will undergo surgery during the week of planning and} \\ \quad \text{no autologous blood donation is planned} \\ 0 \text{ otherwise.} \end{cases}$$

Variable f_{ktp} measures the overtime on operating room $k \in K$ on day $t \in T$ for time slot $p \in P$ and variable η_{kt} the number of medical specialization units assigned to $k \in K$ on day $t \in T$ exceeding the complexity threshold M (see parameters table). Finally, variable ψ_t measures the number of unused anesthetist appointments and variable Φ_t the number of unused blood donation appointments. The complete list of parameters is then reported in Table 1.

3.3.1 Mixed Integer Programming Formulation

In this section we give in detail the MIP formulation split into blocks of constraints starting by the objective function (1), which is rather heterogeneous.

$$MinZ = P_{WT} \left(\sum_{i \in I} \sum_{k \in K} \sum_{t \in T2} \sum_{p=1}^{P} \rho_i \pi_i^{\Delta_i} x_{iktp} + \sum_{i \in I} \rho_i \pi_i^{\Delta_i + 1} z_i \right) + P_{OT} \sum_{k \in K} \sum_{t \in T2} O_{kt} \sum_{p=1}^{P} f_{ktp}$$
$$+ P_{SP} \sum_{k \in K} \sum_{t \in T2} E_{kt} \eta_{kt} + P_V \sum_{i \in I} \varepsilon_i + P_{AV} \sum_{t \in T1} \psi_t + P_{BS} \sum_{i \in I} \zeta_i + P_B \sum_{t \in T1} \Phi_t \qquad (1)$$

Objective function (1) minimizes the waiting time of patients, weighted by priority-based clinical condition coefficients, and the penalty associated with exceeding deadline associated with each patient. The general idea of $\pi_i^{\Delta_i}$ is that the more we get close to the deadline the more it will cost to postpone the surgery of the patient. By means of postponing we define the fact that the patient will not be inserted in the weekly plan. We assume that the function that models the deadline weight can be considered as constant on a weekly basis but exponential from week to week. That means that objective function weights $\pi_i^{\Delta_i}$ and $\pi_i^{\Delta_i+1}$ related to patient planned or unplanned admission in the week of planning can be defined as follows. Let Δ_i be the number of weeks that the patient spent in the waiting list, and DL_i the number of weeks that the patient can wait before reaching the deadline day according to his priority status. Let then define $Delay_i = \Delta_i - DL_i$ the number of elapsed weeks to deadline. In addition, the objective function takes into account the number of non-prosthetic surgeries planned in overtime and the number of different types of prosthetic surgeries, scheduled during the same period in one operating room. Finally, the objective function minimizes the number of unplanned medical anesthetist appointments and blood samples for patients defined as critical or suitable for autologous blood donation. All the above discussed components are weighted by means of suitable coefficients that have been determined through experts feedback and preliminary computational evaluations.

$$\sum_{i \in I_w \cap I_r} x_{iktp} \leq N y_{wktpr} \qquad \forall k \in K; \quad \forall t \in T2; \quad \forall w \in W; \quad \forall p \in P; \quad \forall r \in R_q; \quad \forall q \in Q \tag{2}$$

$$\sum_{i \in I_{Ward} \cap I_r} x_{iktp} \leq N \sum_{w \in W} y_{wktpr} \qquad \forall k \in K; \quad \forall t \in T2; \quad \forall p \in P; \quad \forall r \in R_q; \quad \forall q \in Q \tag{3}$$

Constraints (2) are used to link the binary variable y_{wktpr} to the value assumed by x_{iktp} by ensuring that, if a patient surgery is planned for a specific time slot, then the operating room k is assigned to the corresponding surgeon w on day t, slot p for the type of surgeries related to patient's pathology r. Similarly, constraints (3) ensure that, if the patient belongs to the ward waiting list, he can be assigned to every time slot assigned to every surgeon if it fits with the type of surgeries that will be performed in the time slot.

$$\sum_{k \in K} \sum_{t \in T2} \sum_{p \in P} \sum_{q \in Q} g_{wktpq} \geq L_w \qquad \forall w \in W \tag{4}$$

$$\sum_{k \in K} \sum_{t \in T2} \sum_{p \in P} \sum_{q \in Q} g_{wktpq} \leq U_w \qquad \forall w \in W \tag{5}$$

$$\sum_{i \in Ward} \sum_{k \in K} \sum_{t \in T2} \sum_{p \in P} x_{iktp} \geq L_{ward} \tag{6}$$

$$\sum_{i \in Ward} \sum_{k \in K} \sum_{t \in T2} \sum_{p \in P} x_{iktp} \leq U_{ward} \tag{7}$$

The maximum and minimum number of surgeries that is possible to run during each week is defined by constraints (4) and (5) for each surgeon and by constraints (6) and (7) for the ward depending on department internal organization.

$$\sum_{k\in K}\sum_{t\in T2}\sum_{p\in P} x_{iktp} + z_i = 1 \qquad \forall i \in I \tag{8}$$

Constraints (8) ensure, for each patient, that he is either operated (in a specific day, operating room and time slot) or his surgery is postponed.

$$\sum_{r\in R_q} y_{wktpr} \le S_{ktp} g_{wktpq} \qquad \forall w \in W;\quad \forall k \in K;\quad \forall t \in T2;\quad \forall p \in P;\quad \forall q \in Q \tag{9}$$

$$\sum_{w\in W}\sum_{q\in Q} g_{wktpq} = 1 - a_{ktp} \qquad \forall k \in K;\quad \forall t \in T2;\quad \forall p \in P \tag{10}$$

Constraints (9) activate variable g_{wktpq} as a function of the value assumed by y_{wktpr}. In particular, with constraints (10) we make sure that every operating room, if not assigned to trauma management, must be dedicated to surgeries belonging to the same class of surgeries.

$$\sum_{w\in W}\sum_{r\in R_1} y_{wkt1r} \le M + \eta_{kt} \qquad \forall k \in K;\quad \forall t \in T2 \tag{11}$$

$$\sum_{w\in W}\sum_{r\in R_1} y_{wkt2r} = 0 \qquad \forall k \in K;\quad \forall t \in T2 \tag{12}$$

$$\sum_{i\in I_r} x_{iktp} \le S_{ktp} \qquad \forall k \in K;\quad \forall t \in T2;\quad \forall p \in P;\quad \forall r \in R_1 \tag{13}$$

$$\sum_{i\in I_r} x_{iktp} \le S_{ktp} + f_{ktp} \qquad \forall k \in K;\quad \forall t \in T2;\quad \forall p \in P;\quad \forall r \in R_2 \tag{14}$$

Planning constraints related to prosthetic surgeries characterize constraints (11) and (12). Specifically, in (11) the maximum number of different types of prosthetic surgeries that can be scheduled during the same time slot is defined. Constraints (12) state that prosthetic surgeries can be planned only during morning time slots. Constraints (13) impose that the maximum number of prosthetic surgeries that can be planned in each period is bounded by S_{ktp}, while constraints (14) define the possibility of overtime for non-prosthetic surgeries.

$$\sum_{t\in T1} \varphi_{it} + \varepsilon_i = \mu_i(1 - z_i) \qquad \forall i \in I \tag{15}$$

$$\sum_{k\in K}\sum_{p\in P} \mu_i x_{iktp} \le \sum_{a=1}^{t-h_i-1} \varphi_{ia} + \varepsilon_i \qquad \forall i \in I;\quad \forall t \in T2 \tag{16}$$

$$\sum_{i\in I} \varphi_{it} + \psi_t = C_t \qquad \forall t \in T1 \tag{17}$$

Anesthetist appointment management for planned patients that have been classified as critical is managed by constraints (15)–(17). Specifically, constraints (15) state that if a patient is selected for surgery and he has been classified as critical, his anesthetist appointment can be planned at most in 1 day of the time horizon. Constraints (15) should be considered as soft ones since if there are no anesthetist appointment slots available (17) the appointment will be managed as a pre-surgery hospitalization in order to ease an informal and more direct management of anesthetist evaluation. If an anesthetist consult is planned for the patient, constraints (16) impose that it must be attended at least h_i days before the surgery date. Constraints (17) manage the planning of medical examinations taking into account the number of appointment slots still available.

$$\sum_{t\in T1} \sigma_{it} + \zeta_i = \nu_i(1 - z_i) \qquad \forall i \in I \tag{18}$$

$$\sum_{k\in K}\sum_{p\in P} \nu_i x_{iktp} \le \sum_{a=1}^{t-19} \sigma_{ia} + \zeta_i \qquad \forall i \in I; \quad \forall t \in T2 \tag{19}$$

$$\sum_{t\in T1} \lambda_{it} + \zeta_i = \nu_i(1 - z_i) \qquad \forall i \in I \tag{20}$$

$$\sum_{k\in K}\sum_{p\in P} \nu_i x_{iktp} - \zeta_i \le \sum_{a=t-14}^{t-6} \lambda_{ia} \qquad \forall i \in I; \quad \forall t \in T2 \tag{21}$$

$$\sum_{i\in I} (\sigma_{it} + \lambda_{it}) + \Phi_t = B_t \qquad \forall t \in T1 \tag{22}$$

Similarly, the planning of autologous blood donation is handled by constraints (18)–(22). Namely, constraints (18) and (20) ensure that, if a patient is selected to undergo surgery and he has been defined as suitable for autologous blood donation, both first and second sampling can be planned in at most 1 day of the planning horizon. Constraints (19) ensure that, if the first blood sample is taken, it is planned at least 20 days before the surgery date. As a consequence of first sample planning, constraints (21) ensure that the second blood sample must be executed in a range that goes from 15 to 5 days prior to the surgery date. Constraints (22) manage the planning of blood sampling activities taking into account the number of appointment slots still available. It is important to remark that autologous blood donation is not mandatory, which means that, if missed, it will not be done at a later stage. As a consequence constraints (18)–(22) should be considered as soft ones.

Table 2 Variables definition

$x_{iktp} \in \{0,1\}$	$\forall i \in I,\ \forall t \in T2,\ \forall p \in P,\ \forall k \in K$
$y_{wktpr} \in \{0,1\}$	$\forall w \in W,\ \forall t \in T2,\ \forall p \in P,\ \forall k \in K,\ \forall r \in R_q,\ \forall q \in Q$
$g_{wktpq} \in \{0,1\}$	$\forall w \in W,\ \forall t \in T2,\ \forall p \in P,\ \forall k \in K,\ \forall q \in Q$
$z_i \in \{0,1\}$	$\forall i \in I$
$\varphi_{it} \in \{0,1\}$	$\forall i \in I, \quad \forall t \in T1$
$\psi_t \in \{0,\ldots,C_t\}$	$\forall t \in T1$
$\varepsilon_i \in \{0,1\}$	$\forall i \in I$
$\zeta_i \in \{0,1\}$	$\forall i \in I$
$\sigma_{it} \in \{0,1\}$	$\forall i \in I, \quad \forall t \in T1$
$\lambda_{it} \in \{0,1\}$	$\forall i \in I, \quad \forall t \in T1$
$\Phi_t \in \{0,\ldots,B_t\}$	$\forall t \in T1$
$\eta_{kt} \geq 0$	$\forall t \in T2,\ \forall k \in K$
$f_{ktp} \in \{0,\ldots,F_{ktp}\}$	$\forall t \in T2,\ \forall p \in P,\ \forall k \in K$

$$\sum_{k \in K}\sum_{q \in Q} g_{wktpq} \leq 1 \qquad \forall w \in W; \quad \forall t \in T2; \quad \forall p \in P \tag{23}$$

$$\sum_{k \in K}(y_{w_1ktpr_d} + \sum_{q \in Q}\sum_{r \in R_q} y_{w_2ktpr}) \leq 1 \qquad \forall d \in D : d = \langle w_1, w_2, r_d \rangle; \quad \forall t \in T2; \quad \forall p \in P \tag{24}$$

In constraints (23) we require that each surgeon is not assigned to more than one operating room in the same period. Since it is possible that, for certain types of intervention r_d, some surgeries must be managed together by two surgeons, constraints (24) require that a surgeon w_2 is not assigned to any operating room in the period in which he has to support surgeon w_1.

The overall MIP model is then given by objective function (1) and constraints (2)–(24), plus the variables definition reported in Table 2.

4 A Tool to Support the Planning Activities

The MIP model presented in Sect. 3 has been implemented with IBM ILOG CPLEX Optimization Studio 12.5, and tested with real-world instances provided by a local hospital orthopedic department by using perturbed data for privacy reasons.

The data reported, for each patient, his pseudo ID, priority class, the date of the admission in the waiting list and the type of surgery he has to undergo. In addition to patients basic description, data regarding their clinical status, and the subsequent need to schedule an anesthetist appointment, as well as their suitability to autologous blood donation were defined. We implemented a Microsoft Access database to support input and output data management. The database provides a structured description of real world data and allows a clear representation of proposed optimal schedules. As far as input management concerns the database interface was created in order to simplify the insertion of new patients in the list, the change of patients day availabilities and the analysis of the proposed schedules so as to facilitate the adoption by medical doctors. It is known that a proposed plan not always turns out to be feasible because patients availability is not completely known in advance. After the definition of the subset of patients that could undergo surgery during the week of planning, a phone call to each of them has to be done in order to verify their availability. Since elective patients are not in life-threatening condition they may decide to postpone the surgery. It is then necessary to run again the optimization in order to define a new potential subset of patients that will replace those that are not available. The interface may also be used as a tool for scenario analysis developing a multi-week plan. As a consequence, routines have been implemented in order to remove from waiting lists patients that have been planned in previous weeks.

4.1 Data Analysis and Model Improvement

As already mentioned in Sect. 3, in each period (morning or afternoon sessions), only surgeries related to the same class can be performed. By testing the model with the real-world instances we realized that the optimal solution consisted only of non-prosthetic surgeries. This is due to the low percentage of patients with very high waiting times that need to undergo a prosthetic surgery. Due to organizational goals the department wants to ensure a minimum number of time slots per week dedicated to perform prosthetic surgeries, we then added the following lower bound on the number of time slots that have to be weekly assigned to prosthetic activities

$$\sum_{w \in W} \sum_{k \in K} \sum_{t \in T2} \sum_{p \in P} g_{wktp1} \geq E_{prosthetic} \tag{25}$$

where $q = 1$ is the special case for prosthetic surgeries. A slight strengthening of the model can be obtained by imposing an upper bound on the number of time slots assigned to each surgeon in relation with the number of patients in need of prosthetic surgeries in the surgeon's waiting list. More precisely, the following constraints are imposed

$$\sum_{k \in K} \sum_{t \in T2} \sum_{p \in P} S_{ktp} g_{wktp1} \leq |P_w| \quad \forall w \in W \tag{26}$$

Note that in some cases $|P_w| = 0$ (where $P_w := \{I_w \cap I_r : r \in R_1\}$), thus the effect of constraints (26) is to simplify the model by fixing variables to 0.

4.2 Parameters Tuning and Results Interpretation

The output of the optimization tool consists of two sets of decisions that are related to how OR time slots are assigned to surgeons and which patients have been assigned to OR time slots.

First, we run the MIP model with an initial configuration of equal weights for all objective function components and analyzed the optimized schedule. This experiment shows that with equal weights configurations all possible overtime slots would be used in order to reduce overall patients waiting time,which in most cases, are already missing the priority-driven deadlines. It is clear that the impact of overtime is, in this case, too low if compared to patients long waiting time.

We then decided to tune the weighting coefficients in order to identify the configuration that admits overtime just in case patients deadline is not respected. This coefficient is strongly influenced by the characteristics of the waiting lists. In theory, overtime may be admissible if and only if there is a patient that, if not planned in overtime, will not undergo surgery before his deadline. This means that the definition of the ratio between overtime weight P_{OT} and waiting time weight P_{WT} should be related to the coefficient $\pi_i^{\Delta_i}$ trend line (see Table 1 and [22]) in order to allow overtime only in case a patient missing his deadline. Even if this approach could be reasonable from a theoretical point of view, it is clear that the long waiting times that characterize the patients in the waiting list will lead to the utilization of all available overtime slots as shown in Table 3. More precisely, 16 time slots are available in regular time (2 OR times 2 time slots per day times 5 days minus 4 trauma time slot reservations), 12 of which can be extended to overtime. Table 3 shows that all 12 slots are indeed used in overtime.

By observing the trend line of the waiting factor of Priority A and B patients it is then possible to affirm that an effective overtime penalization can be achieved by setting the overtime weighting factor P_{OT} up to three order of magnitude higher than the waiting time one.

The analysis of model outputs is also interesting in terms of number of time slots dedicated to prosthetic surgeries. It is clear that waiting time imbalance between prosthetic and non-prosthetic patients in waiting lists leads to a number of time slots

Table 3 Results on real-world instances by allowing overtime only for patients missing their deadline

Instance ID	Number of patients	Number of surgeries in overtime	Number of OR time slots	Vars	Cols	CPU time	Obj value
0.1	1,014	12	16	18,587	3,643	135.83	4,261,089.06
0.2	970	12	16	17,759	3,455	155.77	2,001,231.38
0.3	928	12	16	16,635	3,242	84.47	1,322,456.19
0.4	887	12	16	15,773	3,064	81.15	918,003.97

assigned to prosthetic surgeries equal to the lower bound that we impose. This is not true if we apply a rolling planning approach that simulates the impact of the enforcement of the optimal planning over 1 month. In this case, after the third week of planning the number of time slots assigned to prosthetic surgeries increases as a consequence of waiting list cleaning.

4.3 Organizational Change Through Scenario Analysis

The proposed methodology is at the moment not fully comparable with the manual solutions because the current planning is totally informal. Specifically, each surgeon defines his own subset of patients after the time slot assignment. The weekly plan is not made 1 month in advance but in most cases 2 weeks in advance. Anesthetist appointments as well as autologous blood donations are at the moment poorly planned. In terms of anesthetist appointments, there is no fixed number of appointments assigned to the orthopedic department, so, every time a critical patient is planned, the surgeon staff tries to arrange an appointment and, if not possible, plans an early hospitalization for the patient. The short OT notice of planning in some cases forbids an effective scheduling of autologous blood donation. Patients selection follows a qualitative and informal approach in which the waiting list is analyzed and patients with longer waiting times are selected after some considerations based on their priority level. Every time a potential patient is identified, a phone call is made in order to verify his availability on the proposed dates. Finally, it is worth mentioning that the manual solution takes approximately one and a half to 2 h to define the surgical schedules, while the use of the MIP model requires on average five minutes to solve the problem and to propose solutions that can be iteratively adjusted to accommodate specific needs.

To analyze the impact of the various components of the model and in particular of its complex and heterogeneous objective function, we decided to examine how the relaxation or the enforcement of some constraints would affect the optimal proposed solution. We tested four alternative configurations:

1. Optimized configuration (the one discussed so far);
2. Overtime disabled;
3. No bound (4) on the minimum number of time slots for surgeons;
4. Optimized configuration with one additional operating room.

Table 4 shows the results of the comparison of the four configurations. The tests have been made on four instances that have been extracted from the initial waiting list running the optimized configuration over a 1 month planning horizon. In this way the four configurations receive the same input and the objective function value is decreased (see the last but one column in Table 4) because the waiting lists are iteratively emptied over time.

Concerning configurations (1) and (2), a specific comparison of the components of the objective function (not directly shown in Table 4) would allow to observe that,

Table 4 Computational results under different configurations

Instance ID	Number of patients	Number of ORs	Overtime enabled	Bound enabled	CPU time (sec)	Obj value	Variation w.r.t. optimized configuration
0.1	1014	2	1	1	135.83	4,261,089.06	/
0.1	1014	2	0	1	73.84	4,265,938.06	0.114 %
0.1	1014	2	1	0	105.26	4,258,092.06	−0.070 %
0.1	1014	3	1	1	2454.86	4,247,993.06	−0.307 %
0.2	970	2	1	1	155.77	2,001,231.38	/
0.2	970	2	0	1	64.16	2,004,402.39	0.158 %
0.2	970	2	1	0	52.95	1,996,938.39	−0.215 %
0.2	970	3	1	1	646.36	1,991,053.39	−0.509 %
0.3	928	2	1	1	84.47	1,322,456.19	/
0.3	928	2	0	1	57.39	1,324,779.19	0.176 %
0.3	928	2	1	0	91.71	1,319,353.19	−0.235 %
0.3	928	3	1	1	1606.92	1,314,820.94	−0.577 %
0.4	887	2	1	1	81.15	918,003.97	/
0.4	887	2	0	1	63.13	919,619.97	0.176 %
0.4	887	2	1	0	70.65	915,434.97	−0.280 %
0.4	887	3	1	1	202.35	911,884.97	−0.667 %

due to the high volume of patients missing their deadline, the waiting time component is, by far, the most relevant one. The increase (i.e., worsening) in the overall objective function value for all four instances in Table 4 clearly indicates that saving overtime costs is less attractive than reducing patients waiting time. Of course, this is not fully unexpected due of the way the weights in the objective function have been tuned (see, Sect. 4.2). Nevertheless, the observation is definitely structural because in all four instances 12 less patients are planned, i.e., get a surgery, see Table 3. Conversely, the number of potential assignments when no overtime is allowed is reduced, thus the computing time is 35.99 % lower for configuration (2) if compared to the optimized one.

If we then compare configurations (1) and (3) it is possible to observe a slight improvement in terms of objective function solution value just by defining a different subset of time slot assignment to surgeons. That is to say that the actual subdivision of OR time slots to surgeons is not correlated with the real condition of patients waiting in different lists. That is even more important because the improvement in terms of solution quality is reached by planning exactly the same number of patients, exploiting all possible regular and overtime time slots. We can then affirm that the political bounds that are actually implemented in the department lead to slightly suboptimal solutions.

As a last comparison we took the results obtained by configuration (4) evaluating the impact of an increase in the number of operating rooms assigned to the orthopedic department. In this situation it is possible to see that all additional time slots associated with the new operating room would be used, thus increasing the number of patients undergoing surgery in each planning week up to 30 more than the initial configuration. This would result in an improvement in terms of final objective function value related to waiting time reduction. This result is obtained at the price of higher computing times, up to one order of magnitude. However, note that computing times are not currently a crucial discriminating factor. As a final remark it is important to point out that the increased number of operating rooms would lead, due to the high number of patients already missing their deadline, to an increase in terms of overtimes, in the sense that the new operating room would work in overtime, too. As stated for (1)–(2) comparison, the overtime component increase is, for the tested instances, of minor importance, in terms of absolute value, if compared with the waiting time one.

If we focus our attention only on the weekly impact, the differences between the four proposed policies seem to be very little in terms of overall waiting time reduction. That is not the case if we analyze a wider planning horizon. To do that, we defined as a starting point for all configurations the initial waiting list and run, for each of them, a 1 month rolling planning, i.e., for each configuration the input of week $j+1$ in terms of patients in the waiting lists depends on the planning computed by that configuration in week j.

Table 5 shows that, by increasing the time horizon, the differences between the proposed policies clearly emerge. In order to compare the monthly impact of the proposed configurations we report the variation with respect to the optimized configuration in the last column. In addition, columns seven and eight display the impact, in percentage, on the objective function of the components related to scheduled and postponed patients so as to understand the gain in terms of waiting time reduction. The comparison between configurations (1) and (2) shows that the overtime disabling would lead to a performance worsening of 12.88 %, mainly due to the decrease of 46 patients among the planned ones with a resulting increase in terms of overall length of stay in the waiting list.

The impact of the political bounds can be measured evaluating configuration (3). Even if the overall number of patients that undergo surgery in the planning period is very similar, namely 171 for configuration (1) versus 175 for configuration (3), the overall waiting time reduction would lead to a 5.58 % decrease in terms of objective function value. That phenomenon strengthens the perception that political bounds lead to suboptimal solutions because they forbid a planning cycle that follows overall patients needs. If we then evaluate the assignment of an additional operating room to the orthopedic department we can observe that an improvement in terms of objective function value equal to 24.47 % could be reached as a consequence of waiting time reduction. The additional operating room would allow, over a 1 month planning horizon, to plan 112 additional patients, thus reducing the number of them still missing their deadline.

Table 5 Computational results under different configurations over a 4 week planning horizon

Instance ID	Number of patients	Number of ORs	Overtime enabled	Bound enabled	CPU time (sec)	Obj % scheduled (%)	Obj % postponed (%)	Obj value	Variation w.r.t. optimized configuration
0.1	1014	2	1	1	135.83	54.15	45.85	4,261,089.06	
0.2	970	2	1	1	155.77	35.68	64.32	2,001,231.38	
0.3	928	2	1	1	84.47	33.11	66.88	1,322,456.19	
0.4	887	2	1	1	81.15	26.01	73.98	918,003.97	
								8,502,780.60	/
1.1	1014	2	1	0	105.26	56.37	43.63	4,258,092.06	
1.2	970	2	1	0	107.56	44.64	55.36	1,963,624.92	
1.3	926	2	1	0	80.4	35.79	64.21	1,089,167.58	
1.4	883	2	1	0	72.92	35.24	64.75	717,039.59	
								8,027,924.14	–5.58 %
2.1	1014	2	0	1	73.84	47.43	52.57	4,265,938.06	
2.2	982	2	0	1	66.77	27.06	72.94	2,293,036.92	
2.3	950	2	0	1	65.55	25.08	74.92	1,715,845.56	
2.4	918	2	0	1	77.47	22.45	77.55	1,323,454.50	
								9,598,275.04	12.88 %
3.1	1014	3	1	1	2454.86	70.28	29.72	4,247,993.06	
3.2	940	3	1	1	468.38	54.75	45.24	1,289,106.92	
3.3	869	3	1	1	421.83	54.23	45.76	600,397.58	
3.4	798	3	1	1	479.98	57.76	42.23	284,630.58	
								6,422,128.14	–24.47 %

5 Conclusions

Service timeliness, health care appropriateness and resource utilization efficiency are elements of great importance within the Health Sector. The increasing need of virtuous programming practices for health services and the subsequent development of quantitative tools to support decision-making had grown mainly for the set of activities that are considered most critical from the organizational point of view. Operating Theater planning is, among those studied by operations management scientists, the one with the highest systemic impact due to the number and the economic importance of involved resources. Operating rooms are shared and limited resources, therefore they have a strong impact on hospitals both in terms of quality of service provided to patients and of economic and organizational performance.

In this paper we proposed a Mixed Integer Programming model for the Operating Theater planning problem by following two perspectives: (i) we followed the benchmark guidelines defined by Emilia-Romagna Region and (ii) we customized the model to a Local Hospital orthopedic department. The resulting model is an hybridization of literature SCAP models with the inclusion of regional-driven side constraints. Computational experiments suggested that, even if the data we used came from a big regional hospital trust, the problem can be easily solved using a commercial MIP solver. The integration of not yet implemented or poorly managed regional guidelines into the optimization tool emphasizes the importance of the case-study as a driving factor to ease the planning of complex activities. One of the main indications that the optimization tool has given is that some of the constraints that are active in the current planning approach decrease the potential effectiveness of operating room capacity. Another interesting insight for Regional managers is that most of Local Hospitals database and Information Technology (IT) systems do not properly fit to support the quantitative-based planning of OR activities. In fact, ongoing work is currently devoted to integrate some very successful experiences of IT design for OR management, see [23]. Future work is also associated with evaluating the impact of national and regional policies that aim at reducing the number of beds available for each department and, in some cases, to consider shared beds among departments.

Acknowledgments We acknowledge the support of MIUR, Italy, under the grant PRIN 2012. We are indebted to an anonymous referee for a careful reading and insightful suggestions.

References

1. Cardoen, B., Demeulemeester, E., Belin, J.: Operating room planning and scheduling: a literature review. Eur. J. Oper. Res. **201**(3), 921–932 (2010)
2. Guerriero, F., Guido, R.: Operational research in the management of the operating theatre: a survey. Health Care Manag. Sci. **14**(1), 89–114 (2011)
3. Agnetis, A., Coppi, A., Corsini, M., Dellino, G., Meloni, C., Pranzo, M.: Long term evaluation of operating theater planning policies. Oper. Res. Health Care **1**(4), 95–104 (2012)

4. Joustra, P., de Wit, J., Van Dijk, N., Bakker, P.: How to juggle priorities? an interactive tool to provide quantitative support for strategic patient-mix decisions: an ophthalmology case. Health Care Manag. Sci. **14**(4), 348–360 (2011)
5. Chaabane, S., Meskens, N., Guinet, A., Laurent, M.: Comparison of two methods of operating theatre planning: application in belgian hospital. In: 2006 International Conference on Service Systems and Service Management, vol. 1, pp. 386–392 (2006)
6. Guinet, A., Chaabane, S.: Operating theatre planning. Int. J. Prod. Econ. **85**(1), 69–81 (2003)
7. Testi, A., Tanfani, E., Torre, G.: A three-phase approach for operating theatre schedules. Health Care Manag. Sci. **10**(2), 163–172 (2007)
8. Fei, H., Chu, C., Meskens, N., Artiba, A.: Solving surgical cases assignment problem by a branch-and-price approach. Int. J. Prod. Econ. **112**(1), 96–108 (2008) (Special Section on Recent Developments in the Design, Control, Planning and Scheduling of Productive Systems)
9. Conforti, D., Guerriero, F., Guido, R., Cerinic, M., Conforti, M.: An optimal decision making model for supporting week hospital management. Health Care Manag. Sci. **14**(1), 74–88 (2011)
10. Marques, I., Captivo, M., Vaz Pato, M.: An integer programming approach to elective surgery scheduling. OR Spectr. **34**(2), 407–427 (2012)
11. Riise, A., Burke, E.K.: Local search for the surgery admission planning problem. J. Heuristics **17**(4), 389–414 (2011)
12. Adan, I., Vissers, J.: Patient mix optimisation in hospital admission planning: a case study. Int. J. Oper. Prod. Manag. **22**(4), 5–461 (2002)
13. Testi, A., Tanfani, E.: Tactical and operational decisions for operating room planning: efficiency and welfare implications. Health Care Manag. Sci. **12**(4), 363–373 (2009)
14. Roland, B., Di Martinelly, C., Riane, F., Pochet, Y.: Scheduling an operating theatre under human resource constraints. Comput. Ind. Eng. **58**(2), 212–220 (2010)
15. Vissers, J., Adan, I., Bekkers, J.: Patient mix optimization in tactical cardiothoracic surgery planning: a case study. IMA J. Manag. Math. **16**(3), 281–304 (2005)
16. Adan, I., Bekkers, J., Dellaert, N., Vissers, J., Yu, X.: Patient mix optimisation and stochastic resource requirements: a case study in cardiothoracic surgery planning. Health Care Manag. Sci. **12**(2), 129–141 (2009)
17. Denton, B.T., Miller, A.J., Balasubramanian, H.J., Huschka, T.R.: Optimal allocation of surgery blocks to operating rooms under uncertainty. Oper. Res. **58**, 802–816 (2010)
18. Wullink, G., Van Houdenhoven, M., Hans, E.W., van Oostrum, J.M., van der Lans, M., Kazemier, G.: Closing emergency operating rooms improves efficiency. J. Med. Syst. **31**(6), 543–546 (2007)
19. Tanfani, E., Testi, A.: A pre-assignment heuristic algorithm for the master surgical schedule problem (mssp). Ann. Oper. Res. **178**(1), 105–119 (2010)
20. Oostrum, J.M., Houdenhoven, M., Hurink, J., Hans, E., Wullink, G., Kazemier, G.: A master surgical scheduling approach for cyclic scheduling in operating room departments. OR Spectr. **30**(2), 355–374 (2008)
21. Hans, E., Wullink, G., van Houdenhoven, M., Kazemier, G.: Robust surgery loading. Eur. J. Oper. Res. **185**(3), 1038–1050 (2008)
22. Tubertini, P.: Operational Research applied to Regional Healthcare System. Ph.D. Thesis, University of Bologna, Italy (2014)
23. Buccioli, M., Agnoletti, V., Padovani, E., Perger, P.: ehcm: Resources reduction and demand increase, cover the gap by a managerial approach powered by an it solutions. Stud. Health Tech. Inform. **205**, 945–949 (2014)

Recent Issues in International Supply Chain Network Design—Economic Partnership Modeling

Junko Hosoda, Kenichi Funaki and Takafumi Chida

Abstract In this paper, we present a supply chain network design technique with a preferential tariff under an economic partnership agreement. The proposed model has constraints to judge whether a preferential tariff applies under two types of rules which identify the product origin country. This model can estimate the cost reduction amount under preferential tariff utilization, and design the minimum total cost supply chain network. We show results of some case studies of factory location selection under the assumption that applicable new tariff schemes will go into effect and labor cost will rise remarkably in the future. These results suggest two types of strategies: selecting the lowest location and selecting high-cost locations for which a preferential tariff applies. It is confirmed that the proposed model can effectively perform such a quantitative evaluation.

Keywords Supply chain network design · Economic partnership · Preferential tariff · Mixed integer programming

1 Introduction

Previously, one of the advantages to locate manufacturing site in China was lower labor costs. However, since 2000, the labor cost in China has been increasing, and so the role as the 'World's Factory' is moving to Southeast Asia and North America, such as Indonesia and Mexico, from China [1]. The business conditions of supply

J. Hosoda (✉) · K. Funaki · T. Chida
Center for Technology Innovation – Production Engineering Hitachi Ltd, 292, Yoshidacho, Totsukaku, Yokohama 244-0817, Japan
e-mail: junko.hosoda.dp@hitachi.com

K. Funaki
e-mail: kenichi.funaki.bd@hitachi.com

T. Chida
e-mail: takafumi.chida.dg@hitachi.com

K. Fujisawa et al. (eds.), *Optimization in the Real World*, Mathematics for Industry 13, DOI 10.1007/978-4-431-55420-2_8

are always changing in this way. To survive under global competition, it is necessary to establish a supply chain network, which is a series of flows from the procurement of parts to the supply of the product to the market, suitably.

The supply chain network design problem is to determine the location of sites for suppliers, factories, and distribution centers. Between different possible locations, the characteristics, such as labor cost, level of production skill, and shipping cost, differ. For example, some locations where labor cost is low have a high shipping cost because they are far from the market, but another location where labor is expensive might have a low shipping cost because it is near the market. From the many combinations of global location options, we construct the minimum total cost supply chain network.

Graves and Willems have reported a basic framework for supply chain network design [2], for which they provided an optimization model to minimize the total supply chain cost, including the processing cost, trans-it cost, and fixed cost.

With recent globalization, the study of supply chain design from a global point of view has advanced. One of the important issues under a global point of view is the reduction of duty, which is huge in international trade. For example, Arntzen et al. [3] modeled duty drawback and avoidance options.

On the other hand, the expansion of the recent economic partnership has accelerated market opening in various areas around the world. A free trade agreement (FTA) is an agreement to eliminate or lower tariff and trade barriers between two or more countries or regions. Under an FTA, the tariffs on trade between signatories can be lowered or eliminated using a preferential tariff scheme. According to a survey by the Japan External Trade Organization (JETRO), more than 260 FTAs have been made around the world [4].

As such, tariff cost reduction using preferential tariffs is one of the key issues in supply chain strategy, especially in supply chain network design. However, it is necessary to decide the country of origin of the item to export to judge whether a preferential tariff can be applied. The country of origin is judged based on some rules after having clarified the purchase country of raw materials. Therefore, Funaki and Hosoda [5] developed a model having constraints to judge whether a preferential tariff can be applied under one important rule which identifies the product origin country. The purpose of the present paper is to expand that model and develop it into one that can judge whether a preferential tariff can be applied under two rules.

The remainder of this paper is organized as follows. In the next section, we describe the basic setup for the supply chain network design problem. Then, we describe the optimization model with a preferential tariff (Sect. 3). We go on to discuss case studies (Sect. 4) before concluding (Sect. 5).

2 Basic Setup for Supply Chain Network Design Problem

2.1 Definition of Supply Chain Network Design Problem

We model a supply chain as a network $G : (N, A)$ in which nodes in N are functions in the supply chain and arcs in A are links that describe product flow relations among nodes (Fig. 1). Following the network model scheme of Graves and Willems [2], we introduce binary variable z_{nk} for selection of site location for each node. We select one option for each node, i.e. $\sum_{k=1}^{K_n} z_{nk} = 1$, and the values of cost at nodes and links between nodes depend on selection of location options.

Now we formulate the supply chain network design problem as follows:

2.2 Optimization Model

$$\min C = \sum_{n \in N} C1_n \bar{x}_n + \sum_{(m,n) \in A} C2_{m,n} \frac{\beta_m}{\beta_n} \bar{x}_n + \sum_{n \in N} C3_n \tag{1}$$

subject to

$$\bar{x}_n = \beta_n \cdot d \tag{2}$$

$$C1_n = \sum_{k=1}^{K_n} C1_{nk} z_{nk} \tag{3}$$

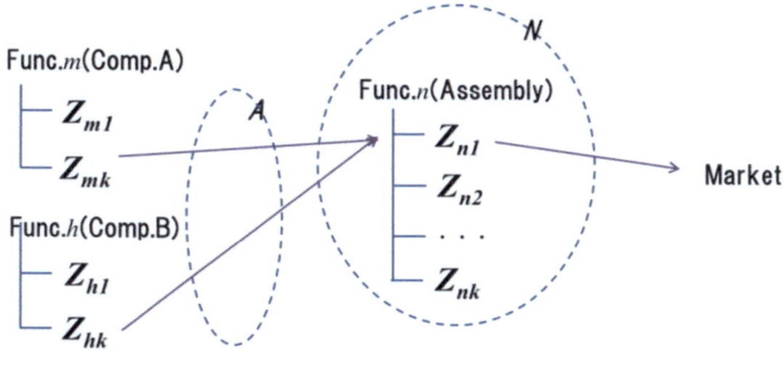

Fig. 1 The schematic model of the supply chain design problem

$$C2_{m,n} = \sum_{h=1}^{K_m} \sum_{k=1}^{K_n} C2_{mh,nk} z_{mh} z_{nk}, \quad (m, n) \in A \tag{4}$$

$$C3_n = \sum_{k=1}^{K_n} C3_{nk} z_{nk} \tag{5}$$

$$\sum_{k=1}^{K_n} z_{nk} = 1 \tag{6}$$

$$z_{nk} \in \{0, 1\}, \forall n, k \tag{7}$$

Here

$G : (N, A)$	Supply chain network consisting of nodes N and arcs A
N	Set of nodes that describe functions in supply chain
A	Set of links that describe product flow relations between nodes
β_n	Quantity of product at node n needed per final product
d	Demand quantity of final product
$\bar{x}_n$	Total quantity of product to be processed at node n
z_{nk}	Binary variable for site location selection, which is 1 if the kth location option for node n is selected and 0 otherwise
K_n	Number of location options for node n
$C1_n$	Unit processing cost at node n
$C1_{nk}$	Unit processing cost at node n with option k
$C2_{m,n}$	Unit transit cost from node m to node n
$C2_{mh,nk}$	Unit transit cost from node m with option h to node n with option k
$C3_n$	Fixed cost at node n
$C3_{nk}$	Fixed cost of node n with option k

The objective function is the sum of total processing cost, total transit cost, and fixed cost incurred depending on the location selection. When a link connects sites in different countries, i.e., describes export from one country to other country, the cost due to tariffs is included in the corresponding transit cost value. The problem is typically a mixed integer programming (MIP) problem and can be solved by standard MIP solvers.

3 Supply Chain Network Design with Preference Tariff

3.1 Scheme of Preferential Tariff

A preferential tariff scheme is applied to products only if they qualify as products originating in an FTA signatory. There are three generally used rules to identify the product origin country [6], as follows.

1. Change in tariff classification (CTC) rule
2. Regional value content (RVC) rule
3. Specified process (SP) rule

3.1.1 Change in Tariff Classification (CTC) Rule

All of the materials, components, and final products have codes called 'tariff classification' codes. According to the CTC rule, in each process (node in our model) occurring in FTA signatories, the tariff classification codes of materials from non-originating countries and the product being processed are compared. If the tariff classification codes of the non-originating materials and the tariff classification of the product differ, this means that the material is being transformed substantially, and so the product will qualify as originating in FTA signatories.

3.1.2 Regional Value Content (RVC) Rule

According to the RVC rule, if the value generated in FTA signatories is a substantial portion of the total value of the product, then the product will qualify as a product originating in FTA signatories.

The RVC value of each product is defined by the following formula:

$$RVC = \frac{V - VNM}{V} \times 100 \tag{8}$$

Here *RVC* is the regional value content value, *V* is the product unit price, and *VNM* is the value of non-signatory-made materials.

Figure 2 shows an example of the RVC calculation. In this figure, the highlighted area corresponds to the FTA area. When component A is supplied from location Z_{mk}, which is a signatory of the FTA, and component B is supplied from locationZ_{hk}, which is not a signatory, then the RVC value at shipping after assembly is calculated as 55 % using Eq. (8).

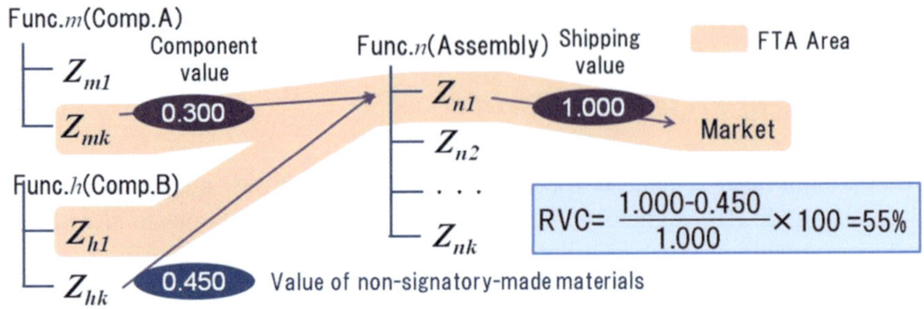

Fig. 2 Example of an RVC calculation

3.1.3 Specified Process (SP) Rule

According to the SP rule, a process in which the product is transformed substantially is specified for each product. If the specified process is in FTA signatories, then the products are considered as originating in FTA signatories.

Each FTA defines one or more rules to be applied. The SP rule is not common to all products, but applies only to the processes for specific product types (e.g., textiles), so in this paper, we focus on the CTC rule and the RVC rule.

3.2 Optimization Model with Preferential Tariff

The basic model to judge whether a preferential tariff applies are described as Eqs. (9)–(12) and we have 4 types of model. First, if the CTC rule needs to be satisfied to obtain qualification as an originating product in FTA signatories, Eq. (13) is applied. Second, if the CTC rule needs to be satisfied, Eqs. (14) and (15) are applied. Third, if both the CTC rule and the RVC rule need to be satisfied, Eqs. (13)–(17) are applied. And finally, if either the CTC rule or the RVC rule needs to be satisfied, Eqs. (9)–(18) are applied.

$$\min C = \sum_{n\in N} C1_n \bar{x}_n + \sum_{(m,n)\in A} \left(C2_{m,n} + C4_{m,n}\right) \frac{\beta_m}{\beta_n} \bar{x}_n + \sum_{n\in N} C3_n \tag{9}$$

subject to Eqs. (2)–(7) and

$$C4_{m,n} = \sum_{h=1}^{K_m} \sum_{k=1}^{K_n} C4_{mh,nk} \left(1 - w_m e_{mh} e_{nk}\right) z_{mh} z_{nk} \; (m, n) \in A \tag{10}$$

$$e_{nk} \in \{0, 1\}, \forall n, k \tag{11}$$

$$w_n \in \{0, 1\}, \forall n \tag{12}$$

Here,

$C4_{m,n}(C4_{mh,nk})$	Unit cost due to a tariff on export from node m (with option h) to node n (with option k)
e_{nk}	Binary parameter to identify locations in the economic partnership region, which is 1 if the kth location option for node n is in the region and 0 otherwise
$w_n\ (w_{n,r})$	Binary variable to judge whether the product at node n qualifies for a preferential tariff, which is 1 if the product at node n satisfies the rules

The Criterion to be used depends on the rule adopted in the agreement. For example, the RVC rule and the CTC rule are used in the North American Free Trade Agreement (NAFTA).

The CTC rule is formulated as follows:

$$e_{nk}\left(1-e_{mh}\right)z_{mh} \le G_{nm} + \left(1-w_{n,CTC}\right) \tag{13}$$

Here,

G_{nm}	Binary parameter to indicate whether the tariff classification code of the product at node n and the tariff classification code of materials at node m are the same or different, which is 1 if both tariff classification cord are same and 0 otherwise

The RVC rule is formulated as follows:

$$\theta \cdot w_{n,RVC} \le \frac{V_n - NV_n}{V_n} \times 100 \tag{14}$$

Here,

θ	A given thresholed value of the FTA
V_n	Value of product at node n
NV_n	Sum of the values of products supplied from the immediate upstream nodes of node n, which are NOT in the economic partnership region.

NV_n is calculated using the following equation:

$$NV_n = \sum_{m \in \breve{N}} \left\{ V_m \sum_{h=1}^{K_m} \left(1-e_{mh}\right) z_{mh} \frac{\beta_m}{\beta_n} \bar{x}_n \right\} \tag{15}$$

If both the CTC rule and the RVC rule need to be satisfied to obtain qualification as an originating product in FTA signatories, the judging process is formulated as follows:

$$w_n \leq w_{n,CTC} \tag{16}$$

$$w_n \leq w_{n,RVC} \tag{17}$$

If either the CTC rule or the RVC rule being satisfied is sufficient to obtain the qualification as an originating product in FTA signatories, the judging process is formulated as

$$w_n \leq w_{n,CTC} + w_{n,RVC} \tag{18}$$

4 Computational Results

In this section, we show the effectiveness of the proposed supply chain design model with a preferential tariff by considering case study results.

In these case studies, based on the 2015 business environment, we select site location options for 2020. Production skill improvement, labor cost increase, and FTA criteria value increase are included in business conditions change expected to occur by 2020 [7]. We choose the minimum cost location options depending on these changes. The following numerical data are artificially generated, but they simulate decision making situations that are increasingly occurring for Japanese companies.

The potential supply chain network comprises the four stages of parts supply, component manufacturing, final product assembly, and markets where the products are sold.

Figure 3 shows site location options in 2015. Because a high level of production skill is necessary for production of part A, there are two location options: Japan and

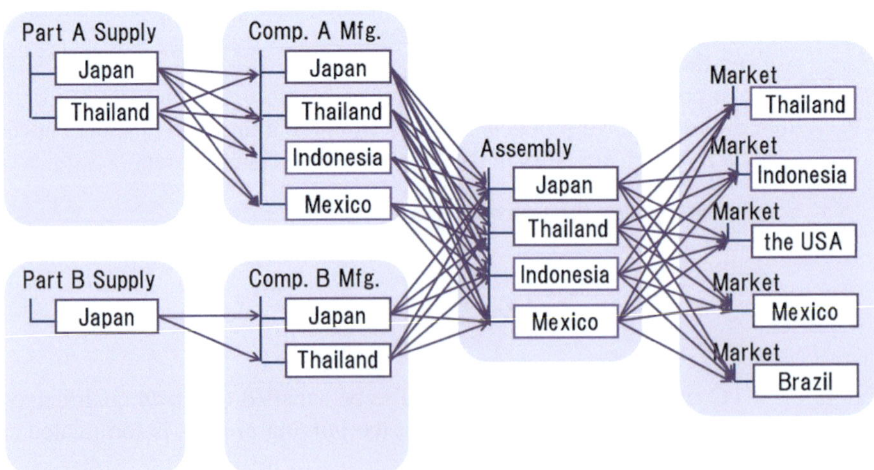

Fig. 3 Site location options in 2015

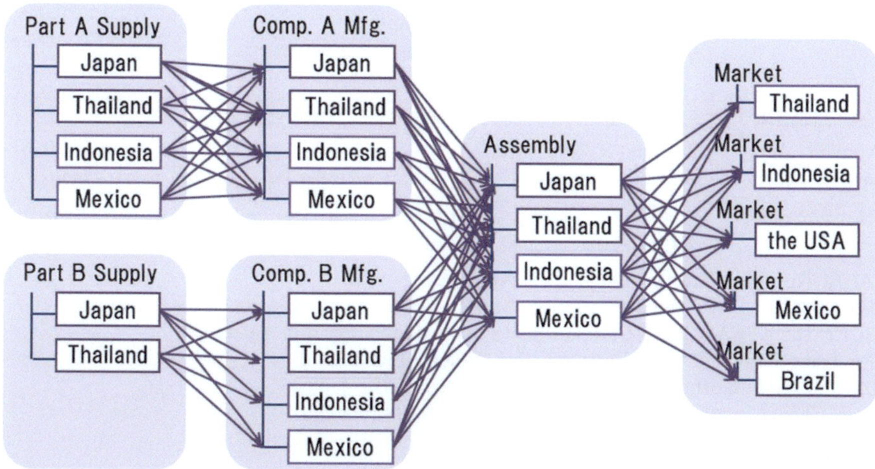

Fig. 4 Site location options in 2020

Thailand. Component A is made of part A and there are four location options: Japan, Thailand, Indonesia, and Mexico. Part B production requires a very high level of production skill, so Part B is supplied from Japan only. Component B made of part B requires a high level of production skill and there are two location options: Japan and Thailand. For the assembly, the location options are Japan, Thailand, Indonesia, and Mexico. There are five markets in this area: Thailand, Indonesia, the USA, Mexico, and Brazil.

Figure 4 shows site location options in 2020. Thailand, Mexico, and Indonesia will have improved production skill in 2020, so the two location options of Indonesia and Mexico of part A and component B are added. In addition, part B can be supplied from Thailand.

There are four applicable economic partnerships in this case, which are listed in Table 1. The applicable economic partnership signatories are listed in Table 2.

First, NAFTA is a North American economic partnership agreement, with the applicable signatories being Mexico and the USA. NAFTA has two rules to identify product origin: a CTC rule and the RVC rule. As the CTC rule, NAFTA applies a change in tariff heading (CTH) rule which requires that all non-originating materials used in the production of the product have undergone a change in tariff classification at the 4-digit level. Under NAFTA, the RVC criterion value is 62.5 %.

Second, the MERCOSUR-Mexico Economic Complementation Agreement (MERCOSUR-Mexico) is a South America and Mexico economic partnership agreement, with the applicable signatories being Mexico and Brazil. The rule to identify product origin is the RVC rule only. Under MERCOSUR-Mexico, the RVC criterion value in Mexico is 35.0 % in 2015, but rises to 40 % in 2020.

Third, the ASEAN-Japan Comprehensive Economic Partnership Agreement (AJCEP) is an ASEAN and Japan economic partnership agreement, with the

Table 1 Summary of applicable economic partnerships

	NAFTA	MERCOSUR-Mexico	AJCEP	TPP
Year effective	2015/2020	2015/2020	2015/2020	2020
Applied rule	CTC and RVC	RVC	CTC or RVC	CTC and RVC
CTC criteria	CTH (2015/2020)	–	CTH (2015/2020)	CTSH (2020)
RVC criteria	62.5 % (2015/2020)	35.0 % (2015) 40.0 % (2020)	40.0 % (2015/2020)	45.0 % (2020)

NAFTA The North American Free Trade Agreement
MERCOSUR-Mexico The MERCOSUR-Mexico Economic Complementation Agreement
AJCEP The ASEAN-Japan Comprehensive Economic Partnership Agreement
TPP Trans-Pacific Strategic Economic Partnership Agreement
CTC Change in tariff classification, *CTH* change in tariff heading
CTSH Change in tariff sub-heading, *RVC* regional value content

Table 2 Applicable economic partnership signatories

	NAFTA	MERCOSUR-Mexico	AJCEP	TPP
Thailand			×	×
Indonesia			×	
Japan			×	×
USA	×			×
Mexico	×	×		×
Brazil		×		

applicable signatories being Thailand, Indonesia, and Japan. The rule is that either a CTC rule or the RVC rule be satisfied to obtain qualification as a product originating in AJCEP signatories. As the CTC rule, AJCEP applies the CTH rule. Under AJCEP, the RVC criterion value is 40.0 %.

Finally, the Trans-Pacific Strategic Economic Partnership Agreement (TPP) is currently under discussion, but it is expected to include the USA, Japan, Thailand and Mexico by 2020 as the applicable countries. We assume that TPP has two rules to identify product origin: a CTC rule and the RVC rule. As the CTC rule, TPP applies the change in tariff sub-heading (CTSH) rule which requires that all non-originating materials used in the production of the product have undergone a change in tariff classification at the 6-digit level. Under TPP, the RVC criterion value is 45.0 %.

Table 3 shows tariff classification of parts, components and final products. Under the CTH rule, part A and component A have the same tariff classification, and part B and component B have the same tariff classification. Under CTSH rule, all items have different tariff classifications.

For the situation described above, we would like to design the best supply chain network, which means choosing the combination of utilized site taking into account the tariff schemes which will apply for each market so as to minimize the total supply chain cost.

Table 3 Tariff classification

	Part A	Part B	Comp. A	Comp. B	Final prod.
CTH	CthA	CthB	CthA	CthB	CthC
CTSH	CtshA	CtshB	CtshC	CtshD	CtshE

Table 4 Parts, components, and final product values (purchase cost $C1_{nk}$) in 2015

	Japan	Thailand	Indonesia	Mexico
Part A	0.12	0.10	0.09	0.10
Part B	0.15	0.12	0.11	0.12
Component A	0.27	0.22	0.20	0.22
Component B	0.36	0.30	0.27	0.30
Final product	1.00	1.00	1.00	1.00

Table 5 Labor costs of components and final product ($C1_{nk}$) in 2015

	Japan	Thailand	Indonesia	Mexico
Component A	0.15	0.12	0.11	0.12
Component B	0.21	0.17	0.16	0.17
Final product	0.37	0.30	0.28	0.30

Table 6 Labor cost ratio in 2020 (Ratio to the cost in Japan)

	Japan	Thailand	Indonesia	Mexico
IDN<MXC	1.00	0.82	0.75	0.82
IDN=MXC	1.00	0.96	0.94	0.94
IDN>MXC	1.00	0.96	0.95	0.86

The numerical details are as follows. The parts, components, and final product values in 2015 are shown in Table 4. Labor cost of components manufacturing and final product assembly in 2015 are shown in Table 5. Table 6 shows our predictions regarding labor cost increases in 2020. This is shown as ratio to the labor cost in Japan.

Table 7 shows the default tariff rate which is applied to items non-originating in FTA signatories. Under FTA scheme, duty-free treatment is applied to items originating in FTA signatories.

The results are as follows for optimizing the supply chain network in five cases. First is a baseline case in 2015 (Baseline). Other cases are supply chain networks designed for 2020 under various business conditions (Table 8). In these cases the labor cost ratio is an especially important parameter. In Cases 1, 2, and 3, the labor cost in Indonesia is assumed to be lower, equal, and or higher than the labor cost ratio in Mexico. Another important parameter is the applicable FTA, and so we also designed Case 4, which is the same as Case 3 expect that TPP is applicable.

Table 7 Default tariff rate ($C4_{mh,nk}$)

	Part A (%)	Part B (%)	Comp. A (%)	Comp. B (%)	Final Prod. (%)
Thailand	80	10	80	10	80
Indonesia	40	5	40	5	10
Japan	0	0	0	0	0
USA	–	–	–	–	2.6
Mexico	0	0	0	0	20
Brazil	35	18	35	18	35

Table 8 Business condition of each case

	Baseline	Case 1	Case 2	Case 3	Case 4
Year	2015	2020	2020	2020	2020
Labor cost ratio	Table 5	IDN < MXC	IDN = MXC	IDN > MXC	IDN > MXC
TPP applicable	No	No	No	No	Yes

We will next explain the optimal supply chain network for each case.

(1) Baseline (2015/Table5/TPP not applicable)

In this case, as a baseline, we designed what supply chain network is optimal in 2015. The optimal supply chain network is shown in Fig. 5. A final product for which a preferential tariff applies under MERCOSUR-Mexico is supplied from Mexico to Mexico and Brazil, because the default tariff in Mexico (20 %) and Brazil (35 %) are higher than the difference between the labor costs in Mexico and Indonesia (10 %). On the other hand, the final product for which a preferential tariff does not apply

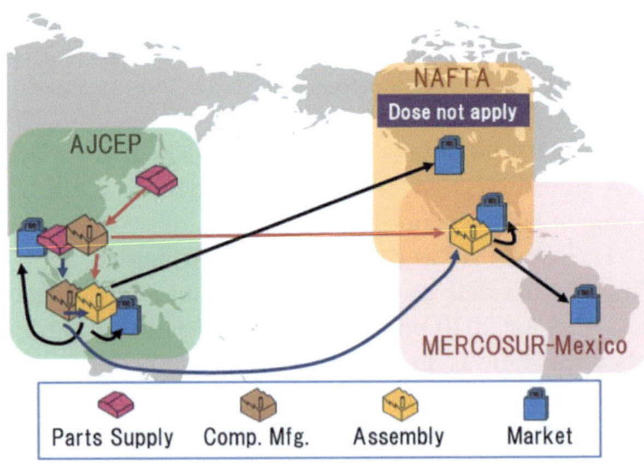

Fig. 5 Optimal supply chain network in Baseline

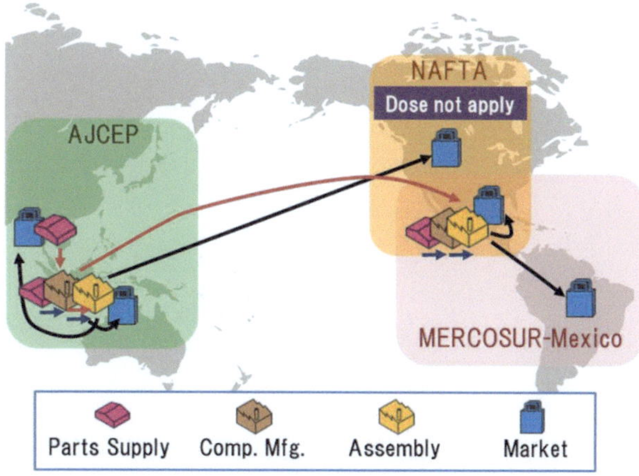

Fig. 6 Optimal supply chain network in Case 1

under NAFTA is supplied from Indonesia to the USA, because the default tariff in the USA (2.6 %) is less than difference between the labor costs in Mexico and Indonesia.

The RVC value under MERCOSUR-Mexico in Mexico is calculated as follows:

$$RVC = \frac{1.00 - (0.36 + 0.27)}{1.00} \times 100 = 37\,\% \tag{19}$$

The RVC criteria value under MERCOSUR-Mexico is 35 %, so the final product in Mexico qualifies as originating in Mexico.

In the AJCEP area, all parts supply and production are carried out in the closed area under which the AJCEP preferential tariff applies.

(2) Case 1 (2020/IDN < MXC/TPP not applicable)

In this case, we designed the optimal supply chain network in 2020 under business condition 1, and the result is shown in Fig. 6. A final product for which a preferential tariff applies under MERCOSUR-Mexico is supplied from Mexico to Mexico and Brazil for a same reason as for Baseline. The difference between Baseline and Case 1 is the production site location of component B. This difference is due to an increase the RVC value in Mexico. Specifically, the RVC criteria value under MERCOSUR-Mexico increases from 35 % in 2015 to 40 % in 2020.

The RVC value under MERCOSUR-Mexico in Mexico is calculated as follows:

$$RVC = \frac{1.00 - 0.36}{1.00} \times 100 = 64\,\% \tag{20}$$

(3) Case 2 (2020/IDN = MXC/TPP not applicable)

In this case, we designed the optimal supply chain network in 2020 under business condition 2, and the result is shown in Fig. 7.

The final product is imported to the USA and a preferential tariff applies under NAFTA. This is because production in Mexico is a priority, due to the low labor cost.

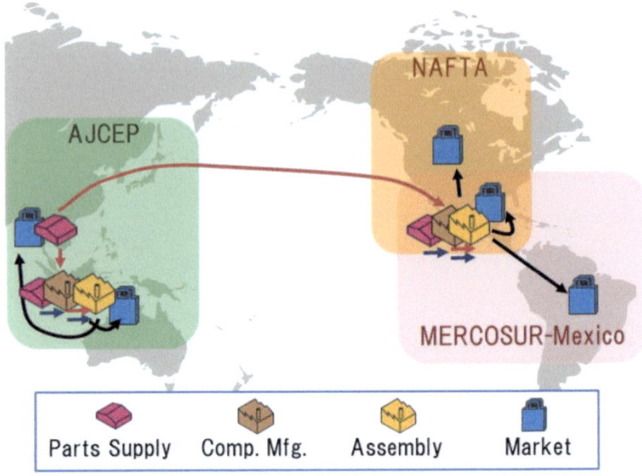

Fig. 7 Optimal supply chain network in Case 2

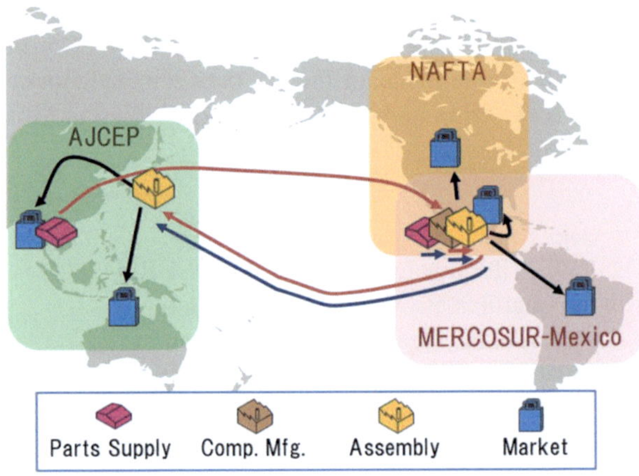

Fig. 8 Optimal supply chain network in Case 3

Part A is supplied from Japan (a non-signatory to NAFTA) to Mexico. The tariff classificasion of the final product and part A are different, so the CTC rule under NAFTA is satisfied. In addition, the RVC value under NAFTA in Mexico is calculated as follows:

$$RVC = \frac{1.00 - 0.15}{1.00} \times 100 = 85\,\% \tag{21}$$

The RVC criteria value under NAFTA is 62.5 %, so the final product in Mexico qualifies as originating in Mexico.

(4) Case 3 (2020/IDN > MXC/TPP not applicable)

In this case, we designed the optimal supply chain network in 2020 under business condition 3, and the result is shown in Fig. 8.

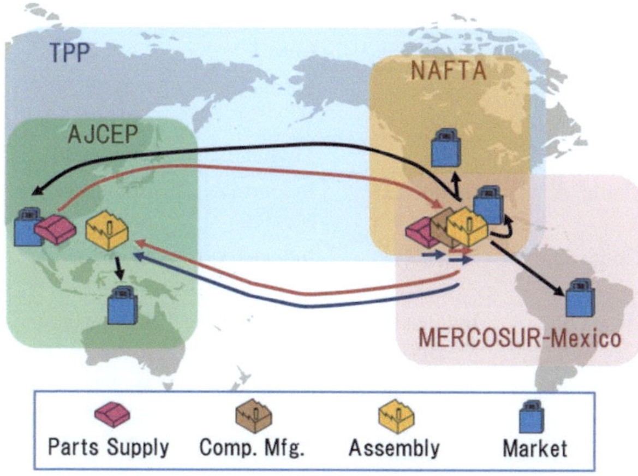

Fig. 9 Optimal supply chain network in Case 4

The final product is supplied from Japan to Thailand and Indonesia. The labor cost is lower in Indonesia (0.35) than in Japan (0.37), but the default tariff of component A in Indonesia (40 %) is higher than that in Japan (duty-free). This is why an assembly site in Japan is selected.

The RVC value under AJCEP in Japan is calculated as follows:

$$RVC = \frac{1.00 - (0.36 + 0.27)}{1.00} \times 100 = 37\,\% \qquad (22)$$

The RVC criteria value under AJCEP is 40 %, so the final product in Japan does not qualify as originating in Japan, but it does qualify under AJCEP by satisfying the CTH rule, despite not satisfy the RVC rule. The tariff clasification of component A and B are different from the tariff classifications of the final product, so the final product supplied from Japan obtains a preferential tariff under AJCEP. Component B might be supplied from Indonesia, if it is supposed that the product is required to satisfy the RVC rule for qualifying under AJCEP.

(5) Case 4 (2020/IDN > MXC/TPP applicable)

Finally, we designed the optimal supply chain network in 2020 under business condition 4, and the result is shown in Fig. 9.

A final product for which a preferential tariff applies under TPP is supplied from Mexico to Thailand. On the other hand, a final product for which a preferential tariff applies under AJCEP is supplied from Thailand to Mexico. An assembly site in Indonesia is not selected, same as in Case 3.

The optimal site locations for each case are shown in Table 9 and the average costs and average tariffs are shown in Table 10, where the average cost is average of the production cost (parts parches, component manufacturing, and final product assembly) for five markets and the tariff is the average of the tariffs (parts, components and final product) for five markets.

Table 9 Optimal site locations (z_{nk}) by case

	Baseline	Case 1	Case 2	Case 3	Case 4
Part A	Indonesia	Indonesia	Indonesia Mexico	Mexico	Mexico
Part B	Japan	Thailand	Thailand	Thailand	Thailand
Comp. A	Indonesia	Indonesia Mexico	Indonesia Mexico	Mexico	Mexico
Comp. B	Indonesia	Indonesia Mexico	Indonesia Mexico	Indonesia Mexico	Thailand Mexico
Assembly	Indonesia Mexico	Indonesia Mexico	Indonesia Mexico	Japan Mexico	Mexico
Applied FTA	M.M. AJCEP	M.M. AJCEP	NAFTA M.M. AJCEP	NAFTA M.M. AJCEP	NAFTA M.M. AJCEP TPP

M.M. MERCOSUR-Mexico

Table 10 Average cost and tariff by case

	Baseline	Case 1	Case 2	Case 3	Case 4
Average cost	0.55	0.53	0.65	0.62	0.60
Average tariff	0.01	0.01	0.00	0.00	0.00

From these results, we can identify two types of strategies: selecting the lowest location (Case 1 and 2) and selecting high-cost locations so as to receive a preferential tariff (Case 3 and 4). We should accept the default tariff if satisfying the rules of origin for the preferential tariff is excessively costly. On the other hand, if the benefit of the preferential tariff is large, we should select high-cost locations so as to satisfy the rules of origin. The strategy changes if the labor cost in Indonesia became more than 4.6 % higher than the labor cost in Mexico.

5 Conclusions

In this paper, we present a supply chain network design techniques with a preferential tariff from an economic partnership agreement. The proposed model has constraints to judge whether a preferential tariff applies under two types of rules for identifying the product origin country. The first type of rules is the CTC rule, which requires that the tariff classification code of non-originating materials and the tariff classification of the product differ. The other type of rule is the RVC rule, which requires that the value added to the product in signatories satisfy an RVC criterion. We modeled these constraints by using MIP.

We show case study results of factory location selection assuming applicable new tariff schemes will go into effect and that the labor cost will rise remarkably in the future. From these results, we can identify two types of strategies: selecting the lowest location and selecting a high-cost location to obtain a preferential tariff. The strategy changes if the labor cost in Indonesia became more than 4.6 % higher than the labor cost in Mexico. These results confirm that our model is effective for performing a quantitative evaluation.

References

1. Derek, M.T., John, D.C.: Global Manufacturing: the End of China's Golden Age? The Boston Company (2014)
2. Graves, S.C., Willems, S.P.: Optimizing the supply chain configuration for new product. Manage. Sci. **51**(8), 1165–1180 (2005)
3. Arntzen, B.C., et al.: Global supply chain management at digital equipment corporation. Interfaces **25**, 69–93 (1995)
4. JETRO List of the World and Japanese FTA, JETRO Report [published in Japanese], Nov 2014
5. Funaki, K., Hosoda, J.: Supply chain network design with preferential tariff under economic partnership agreement. In: Proceeding of the 2014 International Conference on IEOM, pp.1268–1272 (2014)
6. Estevadeordal, A., Suominen, K.: Rules of origin in the world trading system. Paper prepared for the seminar on regional trade agreements and the WTO, the world trade organaization, Nov 2003
7. Sirkin, H.L., Zinser, M., Rose, J.R.: The Shifting Economics of Global Manufacturing. The Boston Consulting Group Report, Aug 2014

MILP Approaches to Optimal Design and Operation of Distributed Energy Systems

Ryohei Yokoyama and Yuji Shinano

Abstract Energy field is one of the practical areas to which optimization can contribute significantly. In this chapter, the application of mixed-integer linear programming (MILP) approaches to optimal design and operation of distributed energy systems is described. First, the optimal design and operation problems are defined, and relevant previous work is reviewed. Then, an MILP method utilizing the hierarchical relationship between design and operation variables is presented. In the optimal design problem, integer variables are used to express the types, capacities, numbers, operation modes, and on/off states of operation of equipment, and the number of these variables increases with those of equipment and periods for variations in energy demands, and affects the computation efficiency significantly. The presented method can change the enumeration tree for the branching and bounding procedures, and can search the optimal solution very efficiently. Finally, future work in relation to this method is described.

Keywords Energy systems · Optimal design and operation · Mixed-integer linear programming · Branch and bound method · Hierarchical approach

1 Introduction

Our daily life depends on energy significantly. We can enjoy convenient goods and things in our daily life by using energy unconsciously. However, since much energy has been produced from fossil fuels, a large amount of fossil fuels has been consumed, and consequently a large amount of carbon dioxide has been emitted to the environment. Once we obtain convenience by using energy, we cannot reduce energy consumption easily, although we feel fears for exhausting fossil fuels and accelerating

R. Yokoyama (✉)
Department of Mechanical Engineering, Osaka Prefecture University, Sakai, Japan
e-mail: yokoyama@me.osakafu-u.ac.jp

Y. Shinano
Department Optimization, Zuse Institute Berlin, Berlin, Germany
e-mail: shinano@zib.de

K. Fujisawa et al. (eds.), *Optimization in the Real World*,
Mathematics for Industry 13, DOI 10.1007/978-4-431-55420-2_9

global warming. To reduce fossil fuel consumption, the following approaches are conceived:

(a) The final energy consumption should be minimized by ourselves.
(b) When renewable energy can be available, we should utilize it positively.
(c) Efficiencies of energy conversion equipment should be heightened by technological development.
(d) Overall energy conversion efficiencies of energy systems composed of multiple pieces of equipment should be heightened by rational design and operation.

Optimization can contribute to the last approach especially.

In recent years, distributed energy systems such as cogeneration ones have been widespread and diversified, and many types of equipment have been installed into them, which means that many alternatives for system design and operation have arisen. Thus, it has become more and more difficult for designers and operators to determine system specifications and operational strategies properly only with their experiences. In addition, not only reliability in energy supply but also economics, energy saving, and environmental impact have become more and more important criteria for system design and operation, with which designers and operators have been burdened more heavily. For the purpose of assisting designers and operators in system design and operation, therefore, it is necessary to develop a tool for providing rational design and operation solutions flexibly and automatically. Mixed-integer linear programming (MILP) approaches are suitable for optimizing system design and operation, because they lead to natural expression of decision variables.

In this chapter, the application of MILP approaches to optimal design and operation of distributed energy systems is described. First, problem definitions and previous works for optimizing operational strategies as well as design specifications of distributed energy systems are reviewed. Then, the optimal design problem, which takes into account the optimal operational strategies, is defined and an MILP method utilizing the hierarchical relationship between design and operation variables is presented. Finally, future work in relation to this method is described.

2 Optimal Operation and Relevant Previous Work

In the operation of distributed energy systems, the operational strategies including operation modes, on/off states of operation, and load levels of equipment are determined properly for a prescribed short period. The period is discretized into multiple sampling time intervals to consider hourly variations in energy demands. Performance characteristics of equipment, relationships between maximum demands and consumptions of purchased utilities, and energy balance and supply-demand relationships are fundamental constraints to be satisfied. Energy demands are parameters whose values should be given. In practice, however, they cannot be given but have to be predicted based on the measurement. Since there are many alternatives for the operational strategies, the optimization can be used to select the best alternative. In the optimal operation, the operational cost, primary energy consumption, and carbon dioxide emission are typical objective functions to be minimized. In the MILP approach, the operation modes and on/off states of operation are expressed by integer variables, and the load levels by continuous variables.

For the operation of cogeneration systems under certain energy demands, a method based on the MILP has been proposed to operate the systems rationally so that they attain the minimum operational cost for heat and power supply, and its effectiveness has been ascertained by comparing this cost minimizing strategy with conventional electric-/thermal-following ones [9]. In this method, however, dynamic characteristics of equipment are neglected, and the operational strategy of equipment is determined statically and independently at each sampling time interval which is set to take account of variations in energy demands. Therefore, the strategy obtained by this method does not necessarily mean the optimal one for a longer period if dynamic characteristics of equipment are taken into account. Moreover, the method tends to make transition of operation modes and on/off states of equipment sensitive even to small variations in energy demands. These are drawbacks to be overcome for implementing the method into a real-time operational advisory system.

To cope with the aforementioned drawbacks of the conventional optimal operation method, an alternative one has been proposed by incorporating equipment startup and shutdown costs. Although this method also neglects dynamic characteristics of equipment, it considers a coupling of operational strategy at all the sampling time intervals over a period considered, and makes transition of operation modes and on/off states of equipment less sensitive. Therefore, the method is considered one of the simple and effective approaches for a real-time operational advisory system. To efficiently solve the optimal operation problem in consideration of equipment startup and shutdown costs, a solution method has been proposed, where the number of candidates for on/off states of equipment is reduced by using information on upper and lower bounds for the optimal value of operational cost [10]. Another solution method based on the dynamic programming and MILP has been proposed, and has been applied to the determination of the operational strategy of a gas turbine cogeneration system [26].

To cope with the aforementioned drawbacks of the conventional optimal operation method, another alternative one can also be considered by incorporating equipment minimum up and down times in place of equipment startup and shutdown costs. This method also makes transition of operation modes and on/off states of equipment less sensitive, and is suitable for a real-time operational advisory system. The method is widely employed for unit commitment of power generation units [18]. Although constraints for minimum up and down times can be formulated explicitly as quadratic equations [19], they are often treated implicitly. For example, although a solution method based on the Lagrange relaxation and dynamic programming has been proposed, it needs a relatively complex solution procedure and leads only to a feasible solution due to a duality gap, which means that there exists a difference between upper and lower bounds for the objective function [5]. On the other hand, to directly solve the optimal operation problem in consideration of equipment minimum up and down times, the constraints for minimum up and down times have been treated as quadratic equations explicitly, and have been transformed into linear ones by adding new variables and constraints, which leads to the optimal solution by the MILP [23].

The aforementioned methods can be useful tools for operating energy systems if energy demands can be predicted accurately before operation. However, energy demands predicted before operation have some uncertainty, which makes it impossible to predict energy demands accurately. If the operation modes and on/off states of equipment are determined on the assumption that the energy demands predicted before operation are certain, the economic and energy saving characteristics expected before operation may not be attained and the deficit in energy supply may occur in operation. This is because the energy demands which occur in operation differ from those predicted before operation. Therefore, operators should consider that energy demands have some uncertainty, evaluate the robustness in economic and energy saving characteristics against the uncertainty, and operate the systems rationally in consideration of the robustness. Therefore, an optimization method based on the MILP has been proposed to conduct the robust optimal operation of energy systems under uncertain energy demands. The uncertainty in energy demands is expressed by their intervals. The operational strategy is determined to minimize the maximum regret or loss in the operational cost under the uncertainty. In addition, a hierarchical relationship among operation modes and on/off states of equipment, energy demands, and load levels of equipment is taken into account. In addition, the method enables one to incorporate equipment startup and shutdown costs as well as equipment minimum up and down times easily [29].

3 Optimal Design and Relevant Previous Work

In the design of distributed energy systems, the specifications including types, capacities, and numbers of equipment as well as maximum demands of purchased utilities are determined properly along with the operational strategies including operation modes, on/off states of operation, and load levels of equipment for a prescribed long period, typically a year. The long period is discretized into multiple shorter ones to consider seasonal and hourly variations in energy demands. Performance characteristics of equipment, relationships between maximum demands and consumptions of purchased utilities, and energy balance and supply-demand relationships are fundamental constraints to be satisfied. Energy demands are parameters whose values should be given. In practice, however, they cannot be given but have to be estimated based on the experience. Since there are many alternatives for the specifications, the optimization can be used to select the best alternative. In the optimal design, the annual total cost, primary energy consumption, and carbon dioxide emission are typical objective functions to be minimized. In the MILP approach, the types, capacities, numbers, operational modes, and on/off states of operation are expressed by integer variables, and the load levels by continuous variables.

In earlier years, the conventional solution algorithm for the MILP which combines the branch and bound method with the simplex one has not been so efficient, and the optimal design problem has often been treated in consideration of single-period operation [16], or multi-period one for a small number of periods [7], to avoid excessive difficulty of the problem. This is because the number of integer variables increases with those of equipment and periods, and it becomes difficult to obtain the optimal solution in a practical computation time by using the conventional solution algorithm. Afterwards, some efforts have been made to treat the optimal design problem in consideration of multi-period operation for a larger number of periods [11, 25]. Nevertheless, equipment capacities have still been treated as continuous variables, and correspondingly performance characteristics and capital costs of equipment have been assumed to be continuous functions with respect to their capacities. This is because if equipment capacities are treated discretely, the number of integer variables increases drastically, and the problem becomes too difficult to solve. As a result, the treatment of equipment capacities as continuous variables causes discrepancies between equipment capacities optimized and available at the market, and expresses the dependence of performance characteristics and capital costs on capacities with worse approximations.

In recent years, commercial MILP solvers have become more efficient, and many applications to the optimal design of distributed energy systems have been conducted in consideration of multi-period operation for a large number of periods. However, only the types of equipment with fixed capacities have been determined in [3, 22], the types and capacities of equipment have been determined, but the capacities have been treated as continuous variables in [2, 20, 21], and the types and numbers of equipment with fixed capacities have been determined in [4, 13, 14]. In addition, the dependence of performance characteristics of equipment on their capacities or part load levels has not been taken into account in [17, 31]. On the other hand, an optimal design method has been proposed in consideration of discreteness of equipment capacities to resolve the aforementioned insufficiency of equipment models [27]. In this method, a formulation for keeping the number of integer variables as small as possible has been presented to solve the optimal design problem efficiently. However, the aforementioned difficulty in the MILP method still exists essentially. Even commercial MILP solvers which are recently available may not derive the optimal solutions in practical computation times.

Recently, an MILP method utilizing the hierarchical relationship between design and operation variables has been proposed to solve the optimal design problem of energy systems efficiently: At the upper level, the optimal values of design variables are searched by the branch and bound method; At the lower level, the values of operation variables are optimized independently at each period by the branch and bound method under the values of design variables given tentatively during the search at the upper level; Lower bounds for the optimal value of the objective function to be minimized are evaluated, and are utilized for the bounding procedures at both the levels. This method has been implemented into open and commercial MILP solvers. Illustrative and practical case studies on the optimal design of cogeneration systems have been conducted, and the validity and effectiveness of the proposed method have been clarified [30].

4 Hierarchical Approach to Optimal Design

4.1 Definition of Optimization Problem

Some commercial MILP solvers which are recently available can solve large scale problems in practical computation times. However, the MILP problem for the optimal design becomes extremely large scale due to discreteness of equipment capacities when the numbers of types and capacities of equipment, and periods increase. In such a case, even commercial MILP solvers may not derive the optimal solutions in practical computation times. Here, an MILP method utilizing the hierarchical relationship between the design and operation variables is presented [30]. It should be noted that this method can be applicable to any problems with the same form in the other areas.

A detailed form of the optimal design problem under consideration is found in [27], but is omitted here. Its general form is expressed as follows:

$$\left.\begin{array}{ll}\text{min.} & z = f_0(\boldsymbol{\eta}) + \sum_{k=1}^{K} f_k(\boldsymbol{\delta}(k), \boldsymbol{x}(k)) \\ \text{sub. to} & \boldsymbol{g}_0(\boldsymbol{\eta}) \leq \boldsymbol{0} \\ & \boldsymbol{g}_k(\boldsymbol{\eta}, \boldsymbol{\delta}(k), \boldsymbol{x}(k)) \leq \boldsymbol{0} \quad (k = 1, 2, \ldots, K) \\ & \boldsymbol{h}_k(\boldsymbol{\eta}, \boldsymbol{\delta}(k), \boldsymbol{x}(k)) = \boldsymbol{0} \quad (k = 1, 2, \ldots, K) \\ & \boldsymbol{\eta} \in \mathbb{Z}^{n_1} \\ & \boldsymbol{\delta}(k) \in \mathbb{Z}^{n_2} \quad (k = 1, 2, \ldots, K) \\ & \boldsymbol{x}(k) \in \mathbb{R}^{n_3} \quad (k = 1, 2, \ldots, K) \end{array}\right\} \tag{1}$$

where $\boldsymbol{\eta}$ is the vector for n_1 design integer variables, which mean the types, capacities, and numbers of equipment, $\boldsymbol{\delta}(k)$ is the vector for n_2 operation integer variables, which mean the numbers of equipment at the on state, and $\boldsymbol{x}(k)$ is the vector for n_3 operation continuous variables, which mean the load levels or energy flow rates of equipment. K is the number of periods set to consider seasonal and hourly variations in energy demands, and k is the index for the periods. f_0 and f_k denote the terms composed of the design and operation variables, respectively, in the objective function z. $\boldsymbol{g}_0$ is the vector for the inequality constraints which relate design variables. $\boldsymbol{g}_k$ is the vector for the inequality constraints for a part of performance characteristics of equipment which relate design and operation variables. $\boldsymbol{h}_k$ is the vector for the equality constraints for the remaining part of performance characteristics of equipment as well as energy balance and supply-demand relationships which relate design and operation variables.

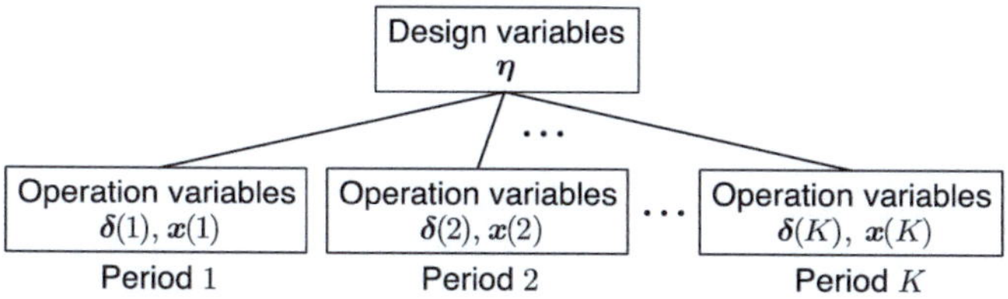

Fig. 1 Hierarchical relationship between design and operation variables

4.2 Basic Concept of Solution

The optimal design problem (1) has the hierarchical relationship between the design and operation variables as shown in Fig. 1. The design variables at the upper level are the integer variables $\boldsymbol{\eta}$, while the operation variables at the lower level are the integer variables $\boldsymbol{\delta}(k)$ and the continuous variables $\boldsymbol{x}(k)$ at each period k. The values of these design and operation variables at all the periods should be optimized simultaneously. However, if the values of the design variables $\boldsymbol{\eta}$ are assumed at the upper level, the constraints $\boldsymbol{g}_k$ and $\boldsymbol{h}_k$ become independent at each period. Thus, the values of the operation variables $\boldsymbol{\delta}(k)$ and $\boldsymbol{x}(k)$ can be determined independently at each period at the lower level. Namely, $\boldsymbol{\eta}$ acts as coupling constraints for all the operation variables. This feature leads to the following hierarchical solution process.

A flow chart for the overall solution process is shown in Fig. 2. In place of the original problem (1), the optimal design and operation problems at the upper and lower levels, respectively, are defined as follows:

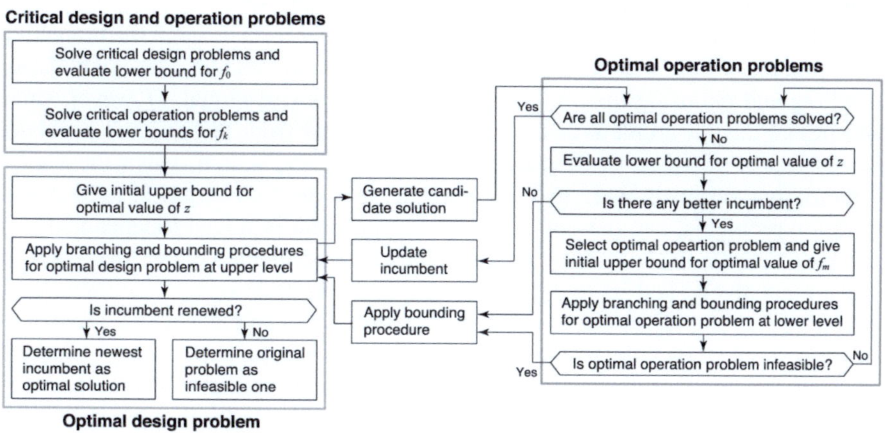

Fig. 2 Overall solution process by hierarchical approach

Optimal design problem at upper level

$$\left.\begin{array}{ll} \text{min.} & z = f_0(\boldsymbol{\eta}) + \sum_{k=1}^{K} f_k(\boldsymbol{\delta}(k), \boldsymbol{x}(k)) \\ \text{sub. to} & \boldsymbol{g}_0(\boldsymbol{\eta}) \leq \mathbf{0} \\ & \boldsymbol{g}_k(\boldsymbol{\eta}, \boldsymbol{\delta}(k), \boldsymbol{x}(k)) \leq \mathbf{0} \quad (k = 1, 2, \ldots, K) \\ & \boldsymbol{h}_k(\boldsymbol{\eta}, \boldsymbol{\delta}(k), \boldsymbol{x}(k)) = \mathbf{0} \quad (k = 1, 2, \ldots, K) \\ & \boldsymbol{\eta} \in \mathbb{Z}^{n_1} \\ & \boldsymbol{\delta}(k) \in \mathbb{Z}^{n_2} \quad (k = 1, 2, \ldots, K) \\ & \boldsymbol{x}(k) \in \mathbb{R}^{n_3} \quad (k = 1, 2, \ldots, K) \end{array}\right\} \tag{2}$$

where $\boldsymbol{\delta}(k)$ is relaxed to continuous variables.

Optimal operation problems at lower level

$$\left.\begin{array}{ll} \text{min.} & f_k(\boldsymbol{\delta}(k), \boldsymbol{x}(k)) \\ \text{sub. to} & \boldsymbol{g}_k(\boldsymbol{\eta}, \boldsymbol{\delta}(k), \boldsymbol{x}(k)) \leq \mathbf{0} \\ & \boldsymbol{h}_k(\boldsymbol{\eta}, \boldsymbol{\delta}(k), \boldsymbol{x}(k)) = \mathbf{0} \\ & \boldsymbol{\delta}(k) \in \mathbb{Z}^{n_2} \\ & \boldsymbol{x}(k) \in \mathbb{R}^{n_3} \end{array}\right\} \quad (k = 1, 2, \ldots, K) \tag{3}$$

where $\boldsymbol{\eta}$ is given tentatively at the upper level. The optimal design problem at the upper level is defined by relaxing $\boldsymbol{\delta}(k)$ to continuous variables in the original problem (1), while the optimal operation problem at the lower level is defined at each period by adopting f_k as the objective function and giving the values of the design variables $\boldsymbol{\eta}$ tentatively.

An upper bound for the optimal value of z is initially set at a sufficiently large value. The optimal values of $\boldsymbol{\eta}$ are searched at the upper level through the branching and bounding procedures used in the branch and bound method. The methods for searching and choosing the optimal values of $\boldsymbol{\eta}$ are conventional. However, since the optimal design problem at the upper level differs from the original problem, the method is modified as described below. During the search, when the values of the design variables $\boldsymbol{\eta}$ are given tentatively, they are transferred to the optimal operation problems, and the optimal operation problem is solved independently at each period at the lower level by the branch and bound method under the values of $\boldsymbol{\eta}$ given tentatively, and its result is returned to the optimal design problem. If an optimal operation problem at a period is infeasible, for example, because the deficit in energy supply arises, the tentative values of $\boldsymbol{\eta}$ cannot become the optimal ones and are discarded, and thus the bounding procedure is applied between the two levels, or outside of the optimization calculations at the upper and lower levels. Otherwise, i.e., if the optimal operation problems at all the periods are feasible, the optimal values of the operation variables $\boldsymbol{\delta}(k)$ and $\boldsymbol{x}(k)$ are determined, the part of the objective function f_k is evaluated correspondingly, and the value of the objective function z is evaluated by adding f_o based on the tentative values of $\boldsymbol{\eta}$. If the value of z is larger than or equal to an upper bound for the optimal value of z, or the value of z for the

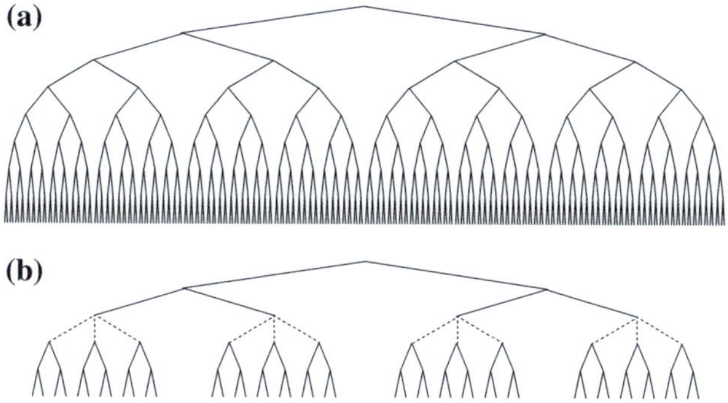

Fig. 3 Enumeration trees for optimization problems. **a** Original problem. **b** Optimal design and operation problems

incumbent obtained previously, the tentative values of η cannot become the optimal ones and are discarded, and thus the bounding procedure is applied between the two levels. Otherwise, the solution corresponding to the tentative values of η becomes a new incumbent, and the previous incumbent is replaced with this new one. When all the branches are searched in the optimal design problem, the incumbent results in the optimal solution of the original problem (1).

Figure 3 shows the enumeration trees for the optimization problems. (a) corresponds to the original problem, while (b) corresponds to the optimal design and operation problems related with each other hierarchically. The number of all the variables in the optimal design problem is the same as that in the original problem (1). However, the number of the integer variables in the optimal design problem is much smaller than that in the original problem. Therefore, the optimal design problem needs a smaller memory size as well as a shorter computation time to apply the branching and bounding procedures in the branch and bound method. In addition, the number of the variables in the optimal operation problem at each period is quite small, and the optimal operation problem can be solved easily. As a result, the proposed method has better features in memory size and computation time as compared with the direct solution of the original problem.

4.3 Evaluation of Lower Bounds

It is possible to solve the optimal design problem at the upper level and the optimal operation problems at the lower level by using the branch and bound method. However, it is suitable to introduce bounding procedures for efficient optimization calculations. Here, lower bounds for parts of the objective function are derived and bounding procedures based on them are applied in consideration of the hierarchical relationship.

Since the objective function is divided into the terms composed of the design and operation variables, lower bounds are also evaluated for the corresponding terms independently. First, the following critical design problem at each period is considered:

$$\left.\begin{array}{ll} \text{min.} & f_0(\boldsymbol{\eta}) \\ \text{sub. to} & \boldsymbol{g}_0(\boldsymbol{\eta}) \leq \boldsymbol{0} \\ & \boldsymbol{g}_k(\boldsymbol{\eta}, \boldsymbol{\delta}(k), \boldsymbol{x}(k)) \leq \boldsymbol{0} \\ & \boldsymbol{h}_k(\boldsymbol{\eta}, \boldsymbol{\delta}(k), \boldsymbol{x}(k)) = \boldsymbol{0} \\ & \boldsymbol{\eta} \in \mathbb{Z}^{n_1} \\ & \boldsymbol{\delta}(k) \in \mathbb{Z}^{n_2} \\ & \boldsymbol{x}(k) \in \mathbb{R}^{n_3} \end{array}\right\} \quad (k = 1, 2, \ldots, K) \tag{4}$$

This problem is obtained by considering the design variables $\boldsymbol{\eta}$ and only the operation variables at the kth period $\boldsymbol{\delta}(k)$and $\boldsymbol{x}(k)$ as well as the term composed of only the design variables in the objective function f_0. This means that the coupling constraints by $\boldsymbol{\eta}$ for the operation variables other than those at the kth period are removed, and that the objective function is adopted only for the design variables. The optimal solution of this problem gives the minimum of f_0 subject to the constraints for the operation variables at the kth period. Thus, the maximum among the minimums of f_0 for all the K critical design problems becomes a lower bound for f_0 in the original problem (1). This lower bound is denoted by $\underset{\sim}{f}{}_0^{\mathrm{D}}$ to show the process of applying bounding procedures below.

Second, the following critical operation problem at each period is considered:

$$\left.\begin{array}{ll} \text{min.} & f_k(\boldsymbol{\delta}(k), \boldsymbol{x}(k)) \\ \text{sub. to} & \boldsymbol{g}_0(\boldsymbol{\eta}) \leq \boldsymbol{0} \\ & \boldsymbol{g}_k(\boldsymbol{\eta}, \boldsymbol{\delta}(k), \boldsymbol{x}(k)) \leq \boldsymbol{0} \\ & \boldsymbol{h}_k(\boldsymbol{\eta}, \boldsymbol{\delta}(k), \boldsymbol{x}(k)) = \boldsymbol{0} \\ & \boldsymbol{\eta} \in \mathbb{Z}^{n_1} \\ & \boldsymbol{\delta}(k) \in \mathbb{Z}^{n_2} \\ & \boldsymbol{x}(k) \in \mathbb{R}^{n_3} \end{array}\right\} \quad (k = 1, 2, \ldots, K) \tag{5}$$

This problem is obtained by replacing f_0 with f_k in the aforementioned critical design problem. This means that the coupling constraints by $\boldsymbol{\eta}$ for the operation variables other than those at the kth period are removed, and that the objective function is adopted only for the operation variables at the kth period. The optimal solution of this problem gives the minimum f_k subject to the constraints for the operation variables at the kth period. Thus, this minimum becomes a lower bound for f_k in the original problem (1). This lower bound is denoted by $\underset{\sim}{f}{}_k^{\mathrm{O}}$ to show the process of applying bounding procedures below.

The number of integer variables in each of these critical design and operation problems is much smaller than that in the original problem (1), and each problem can be solved easily. In addition, if the optimal values of the design variables $\boldsymbol{\eta}$ for all the critical design and operation problems coincide with one another, the obtained solutions give the optimal solution of the original problem. This is because the coupling constraints by $\boldsymbol{\eta}$ removed in the critical design and operation problems are satisfied, and the lower bounds $\underset{\sim}{f}{}_0^{\mathrm{D}}$ and $\underset{\sim}{f}{}_k^{\mathrm{O}}$ become the optimal values of f_0 and f_k, respectively.

4.4 Bounding Operation

4.4.1 Optimal Design Problem

As aforementioned, the optimal design problem at the upper level (2) is considered in place of the original problem (1) by relaxing the integer operation variables $\boldsymbol{\delta}(k)$ to continuous ones. The optimal values of the design variables $\boldsymbol{\eta}$ are searched by the branch and bound method. During this search, the bounding procedure is applied at each branching node at the upper level by using a conventional method. The continuous relaxation problem corresponding to each branching node is usually solved, and the value of the objective function z for its optimal solution is used as a lower bound for the bounding procedure. In the problem under consideration, this value can be divided into those for the design and operation variables. Thus, these values are also used as lower bounds for the optimal values of f_0 and f_k. These lower bounds are denoted by $\underset{\sim}{f}{}_0^{\mathrm{C}}$ and $\underset{\sim}{f}{}_k^{\mathrm{C}}$. On the other hand, the aforementioned lower bounds and $\underset{\sim}{f}{}_0^{\mathrm{D}}$ and $\underset{\sim}{f}{}_k^{\mathrm{O}}$ are used at all the branching nodes. Therefore, more effective lower bounds for the optimal values of f_0 and f_k can be selected from the two ones. As a result, a lower bound for the optimal value of the objective function z is evaluated as follows:

$$\underset{\sim}{z} = \max(\underset{\sim}{f}{}_0^{\mathrm{C}}, \underset{\sim}{f}{}_0^{\mathrm{D}}) + \sum_{k=1}^{K} \max(\underset{\sim}{f}{}_k^{\mathrm{C}}, \underset{\sim}{f}{}_k^{\mathrm{O}}) \tag{6}$$

If this value is larger than or equal to an upper bound for the optimal value of z, or the value of z for the incumbent $\tilde{z}$, the bounding procedure is applied at the corresponding branching node at the upper level. Otherwise, the branching procedure is further continued.

4.4.2 Optimal Operation Problem

The optimal operation problem at each period at the lower level (3) is considered by giving the values of the design variables $\boldsymbol{\eta}$ tentatively. Each optimal operation problem can be solved independently from the other ones. This is because the values of the design variables $\boldsymbol{\eta}$ are given tentatively, and the coupling constraints by $\boldsymbol{\eta}$ are removed. Thus, the optimal operation problems are solved sequentially. Here, the following bounding procedure between the two levels can be applied. Before solving each optimal operation problem in the specified order, a lower bound for the optimal value of the objective function z is evaluated as follows:

$$\underset{\sim}{z} = f_0 + \sum_{k\in S} f_k + \sum_{k\in U} \max(\underset{\sim}{f}_k^{\mathrm{C}}, \underset{\sim}{f}_k^{\mathrm{O}}) \tag{7}$$

where S and U are the sets for the solved and unsolved optimal operation problems, respectively. This equation means that the first term in the right hand side is calculated from the values of the design variables $\boldsymbol{\eta}$, that the second term is calculated from the optimal values of the objective functions f_k for the solved optimal operation problems, and that the third term is calculated from the lower bounds for the unsolved optimal operation problems, where $\underset{\sim}{f}_k^{\mathrm{C}}$ is a lower bound for the optimal value of f_k evaluated by the continuous relaxation problem after the values of all the design variables are given tentatively. If this value is larger than or equal to an upper bound for the optimal value of z, or the value of z for the incumbent $\tilde{z}$, it is judged that the values of the design variables cannot give the optimal solution of the original problem. Thus, the bounding procedure is applied between the two levels without solving the unsolved optimal operation problems, and this information is transferred to the optimal design problem. Else if the optimal operation problem is infeasible, the bounding procedure is applied between the two levels without solving the unsolved optimal operation problems, and this information is also transferred to the optimal design problem. Otherwise, the next optimal operation problem is solved. This same procedure is repeated until the last optimal operation problem is solved. Since the value of $\max(\underset{\sim}{f}_k^{\mathrm{C}}, \underset{\sim}{f}_k^{\mathrm{O}})$ is replaced with the optimal value of f_k after the kth optimal operation problem is solved, the lower bound $\underset{\sim}{z}$ increases as the optimal operation problems are solved sequentially, which increases the possibility of applying the bounding procedure before solving the following optimal operation problems.

When each optimal operation problem is solved by the branch and bound method, the bounding procedure is applied at each branching node at the lower level by using the conventional method. The continuous relaxation problem corresponding to each branching node is usually solved as aforementioned. To apply this bounding procedure more efficiently during the solution of an optimal operation problem, say, the mth one, an initial upper bound for the optimal value of the objective function f_m is evaluated as follows:

$$\tilde{f}_m = \tilde{z} - \left(f_0 + \sum_{k\in S} f_k + \sum_{k\in U\backslash\{m\}} \max(\underset{\sim}{f}_k^{\mathrm{C}}, \underset{\sim}{f}_k^{\mathrm{O}}) \right) \tag{8}$$

If this initial upper bound is not replaced with the value of f_m for an incumbent, it is judged that the solution of the mth optimal operation problem is not necessary. Thus, the bounding procedure is applied between the two levels without solving the unsolved optimal operation problems, and this information is transferred to the optimal design problem.

4.5 Implementation into Solvers

The presented method is implemented into a commercial MILP solver IBM ILOG CPLEX Optimization Studio V12.5.1 [8]. Since this MILP solver is one of the powerful commercial solvers using the linear programming based branch and cut method, it is applicable to large scale problems with large numbers of variables. Here, this MILP solver can be used to solve the optimal design problem at the upper level as well as the critical optimal operation problems. Since the optimal operation problems at the lower level are small scale, an open MILP solver published in [12] is used to solve them. In addition, some callback functions for CPLEX are utilized to control the flow of the optimization calculations for the optimal design and operation problems. They are called by the solver for the optimal design problem and exchanges information between the optimal design and operation problems.

The evaluation of the lower bound by Eq. (7) and the bounding procedure based on it are incorporated into the program for controlling the flow of all the optimization calculations. However, the evaluation of the lower bound by Eq. (6) and the bounding procedure based on it are not incorporated into the commercial solver for the optimal design problem, because CPLEX does not allow them. In place of them, therefore, the evaluation of the lower bound by Eq. (7) and the bounding procedure based on it with $S = \varnothing$, or without solving any optimal operation problems are also used. Thus, the critical design problems do not need to be solved. The evaluation of the upper bound by Eq. (8) is incorporated into the open solver for the optimal operation problems.

In order to use CPLEX as a solver for the optimal design problem at the upper level, the following three features have to be implemented:

(a) Reject all the solutions found in CPLEX not to update the incumbent at the upper level, although the corresponding values of the design variables are used to solve the optimal operation problems at the lower level.
(b) Set an upper cutoff value to remove branching nodes with their lower bounds larger than the value of the objective function for a new incumbent, if it is found by solving the optimal operation problems at the lower level.
(c) Prohibit generating the same solutions more than once to find a new incumbent efficiently, that is, generate cuts to reject the solutions.

Features (a) and (b) can be implemented by using the incumbent callback of CPLEX. The incumbent callback is called when an integer solution is found but before this solution replaces the incumbent during the CPLEX solution procedure. It interrupts the solution procedure to reject the solution, invokes a solver for the optimal operation problems at the lower level, sets an upper cutoff value if a new incumbent is found, and continues the solution procedure. Feature (c) can be implemented by using the cut and lazy constraint callbacks of CPLEX. When a solution is found during the CPLEX solution procedure, the corresponding constraint which prohibits generating the same solutions is added in the following solution procedure. This constraint is obtained by expressing all the integer variables with binary ones and using the values of the binary variables corresponding to the solution as shown in [1, 15]. However, it is not allowed to add this constraint in the incumbent callback because of the restriction of the library. Thus, the incumbent is stored in the incumbent callback, while the corresponding constraint is added in the cut or lazy constraint callback.

4.6 Application to Cogeneration System

4.6.1 Outline

As a practical case study, the presented method is applied to the optimal design of a gas turbine cogeneration system for electricity, cold water, and steam supply [27]. The optimal design problem is solved by not only the presented method but also the conventional method, and the results obtained by these methods are compared with each other. Here, the following two cases are investigated by the presented method to show the effect of the bounding procedure based on the lower bound on the computation efficiency:

Case I: Equations (6)–(8) are not taken into account.
Case II: Equations (7) and (8) are taken into account.

Both the design and operation variables are simultaneously optimized by the conventional method using the same commercial solver. Here, the optimization calculation is carried out by setting CPLEX parameter "mipordind" at 1. This parameter is used to assign priorities to integer variables for selecting branching variables. Higher priorities are given to the design integer variables, and lower priorities to the operation integer variables. All the optimization calculations are carried out on a MacBook Air with the Intel Core 2 Duo processor of 1.4 GHz and RAM of 2 GB.

4.6.2 Conditions

Figure 4 shows the super structure for the cogeneration system, which has gas turbine generators, waste heat recovery boilers, gas-fired auxiliary boilers, electric compression refrigerators, steam absorption refrigerators, a receiving device for purchasing electricity, and pumps for supplying cold water. It is assumed that each type of the generators, boilers, and refrigerators has four units with a same capacity, and that the gas turbine generators and waste heat recovery boilers are selected together as cogeneration units.

Table 1 shows the capacities and performance characteristic values of candidates of equipment for selection. Although the performance characteristic values are shown only at the rated load level, their changes at part load levels are also taken into account. In addition to the equipment, the maximum demands of electricity and city gas purchased from outside utility companies are also determined, and are selected among discrete values by 1.0 MW and 0.5×10^3 Nm^3/h, respectively. The capacity of

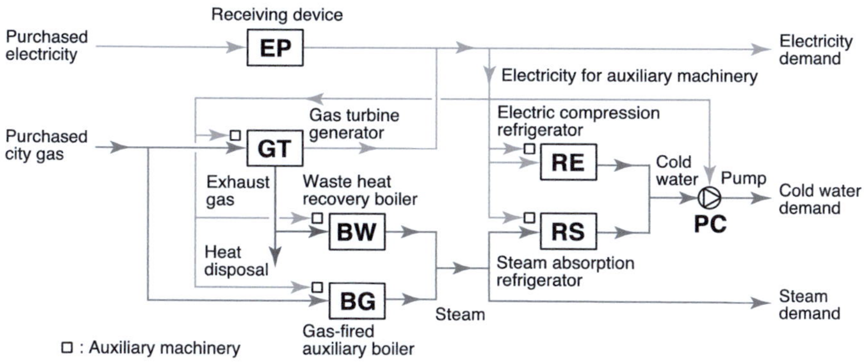

Fig. 4 Configuration of gas turbine cogeneration system

the receiving device for purchasing electricity is also selected among discrete values by 1.0 MW correspondingly. The pumps are common to all the possible structures, and only their power consumption is considered.

Two hotels and four office buildings with the total floor area of $383.7 \times 10^3\,m^2$ in Osaka, Japan are selected as the buildings which are supplied with electricity, cold water, and steam by the cogeneration system. To take seasonal and hourly variations in energy demands into account, a typical year is divided into three representative days in summer, mid-season, and winter whose numbers of days per year are set at 122, 122, and 121 d/y, respectively, and each day is further divided into 24 sampling time intervals of 1 h, respectively. Thus, the year is divided into 72 periods correspondingly.

The annual total cost is adopted as the objective function. In evaluating the annual total cost, the capital recovery factor is set at 0.964 with the interest rate 0.05 and the life of equipment 15 y.

4.6.3 Results

Table 2 shows the optimal values of the design variables obtained by the presented method. Three units of gas turbine cogeneration unit #8 are selected. This is because the power generating and waste heat recovery efficiencies at the rated load level are the second and first highest, respectively, among the ten candidates, and additionally these efficiencies at part load levels are relatively high among the candidates. A unit of electric compression refrigerator #4 is selected. This is because the coefficient of performance at the rated load level is highest among the four candidates. A unit of gas-fired auxiliary boiler #3 and four units of steam absorption refrigerator #2 are selected to supplement steam and cold water supply, respectively.

Table 3 compares the two methods in terms of solution and computation time. The presented method derives the optimal solution in a practical computation time even in case I. The addition of the bounding procedure by the lower bound shortens the computation time in case II. On the other hand, the conventional method also

Table 1 Capacities and performance characteristic values of candidate equipment

Equipment	Capacity/performance[a]	Candidate			
Gas turbine cogeneration unit		#1	#2	#3	#4
	Max. power output MW	1.29	1.60	2.00	2.40
	Max. steam output MW	5.69	3.34	4.10	4.57
	Power generating efficiency	0.140	0.173	0.169	0.179
	Heat recovery efficiency	0.617	0.362	0.347	0.341
		#5	#6	#7	#8
	Max. power output MW	2.93	3.50	3.54	4.36
	Max. steam output MW	6.44	6.97	6.89	8.92
	Power generating efficiency	0.256	0.271	0.273	0.273
	Heat recovery efficiency	0.563	0.540	0.531	0.559
		#9	#10		
	Max. power output MW	5.23	5.32		
	Max. steam output MW	8.91	9.05		
	Power generating efficiency	0.301	0.306		
	Heat recovery efficiency	0.513	0.521		
Gas-fired auxiliary boiler		#1	#2	#3	#4
	Max. steam output MW	5.24	6.55	7.86	9.82
	Thermal efficiency	0.92	0.92	0.92	0.92
Electric compression refrigerator		#1	#2	#3	#4
	Max. cooling output MW	2.82	3.52	4.22	5.28
	Coefficient of performance	4.57	4.73	4.76	5.04
Steam absorption refrigerator		#1	#2	#3	#4
	Max. cooling output MW	3.46	5.18	6.91	8.64
	Coefficient of performance	1.20	1.20	1.20	1.20

[a] At rated load level

Table 2 Optimal values of capacities and numbers of equipment, and maximum demands of utilities

Equipment/utility	Candidate	Number	Capacity
Gas turbine cogeneration unit	#8	3	13.08 MW
Gas-fired auxiliary boiler	#2	1	7.86 MW
Electric compression refrigerator	#1	1	5.28 MW
Steam absorption refrigerator	#3	4	20.72 MW
Receiving device	–	–	4.00 MW
Electricity maximum demand	–	–	4.00 MW
City gas maximum demand	–	–	4.50×10^3 Nm^3/h

derives the optimal solution, but it cannot be ascertained whether the derived solution is optimal or not because the limit for computation memory is attained. In addition, it takes an extremely long computation time.

Table 3 Comparison of conventional and presented methods in terms of solution and computation time

Conventional method		Proposed method		
Solution and objective $\times 10^9$ yen/y	Computation time s	Solution and objective $\times 10^9$ yen/y	Case	Computation time s
Feasible	9167.2[a]	Optimal	I	945.1
1.451		1.451	II	529.4

[a] Attains limit for computation memory

Table 4 Numbers of candidate solutions and optimal operation problems

Number of candidate solutions				Number of optimal operation problems	
All	Removed before lower level	Removed at lower level	Updated incumbents	All	Solved
636	171	463	14	45792	15708

Table 4 shows the numbers of all the candidate solutions which can be generated at the upper level, the candidate solutions removed by the bounding procedures before and while solving the optimal operation problems at the lower level, and the updated incumbents as well as the numbers of all the optimal operation problems and the solved ones. The numbers of all the candidate solutions and optimal operation problems in the first and fifth columns, respectively, are obtained in case I. The numbers of the candidate solutions removed before the lower level, removed at the lower level, and updated incumbents, and the solved optimal operation problems in the second to fourth and sixth columns, respectively, are obtained in case II. This result shows that the bounding procedures before and while solving the optimal operation problems at the lower level remove more than 26 and 72 %, respectively, of the candidate solutions. As a result, the number of the updated incumbents is only less than 3 % of that of the candidate solutions. In addition, the number of the solved optimal operation problems is only less than 35 % of that of all the optimal operation problems.

5 Relevant Future Work

The following future work in relation to the hierarchical approach to the optimal design is under consideration:

(a) Equipment for not only energy conversion but also energy storage may be installed into distributed energy systems. In such a case, the operational strategies should be determined over the period for variations in energy demands and usage of equipment for energy storage. Thus, larger scale critical and optimal operation problems need to be solved. The applicability of the presented method should be investigated.

(b) From the viewpoint of assisting designers, it is important to provide fruitful solution information by obtaining not only the optimal solution but also suboptimal ones which follow the optimal one without any omissions, or K-best solutions [6, 20]. A conventional MILP solver can derive K-best solutions through K optimization calculations by adding the constraints which exclude the solutions found previously [1]. On the other hand, the presented method can derive K-best solutions through an optimization calculation. The effectiveness of the presented method should be investigated.

(c) The performance of distributed energy systems depends on energy demands significantly. However, energy demands must be estimated at the design stage, and consequently they have uncertainty. Although a robust optimal design method based on the minimax regret criterion has been proposed, it can treat only linear models [24]. Thus, it should be extended so that it can treat mixed-integer linear models. The applicability of the presented method to this extension should be investigated.

(d) In utilizing the presented method in practice, it is necessary to generate programs and data for the optimization calculation flexibly and automatically. In addition, it is necessary to prepare a user interface to support it. The presented method should be implemented into the general-purpose optimal planning system for design OPS-Design developed for this purpose [28].

(e) Optimal design problems of various systems may have the same form with those of distributed energy systems. The applicability of the presented method to such problems should be investigated.

Acknowledgments A part of this work has been conducted within the Research Campus Modal funded by the German Federal Ministry of Education and Research (fund number 05M14ZAM) and been supported by the EU COST Action TD 1207. In addition, a part of this work has been supported by the IBM Academic Initiative. The authors would like to thank the anonymous reviewer for his/her valuable comments and suggestions to improve the quality of the paper.

References

1. Balas, E., Jeroslow, R.: Canonical cuts on the unit hypercube. SIAM J. Appl. Math. **23**(1), 61–69 (1972)
2. Buoro, D., Casisi, M., De Nardi, A., Pinamonti, P., Reini, M.: Multicriteria optimization of a distributed energy supply system for an industrial area. Energy **58**, 128–137 (2013)
3. Buoro, D., Casisi, M., Pinamonti, P., Reini, M.: Optimal synthesis and operation of advanced energy supply systems for standard and domotic home. Energy Convers. Manag. **60**, 96–105 (2012)
4. Carvalho, M., Serra, L.M., Lozano, M.A.: Optimal synthesis of trigeneration systems subject to environmental constraints. Energy **36**(6), 3779–3790 (2011)
5. Dotzauer, E.: Algorithms for short-term production-planning of cogeneration plants. Ph.D. Dissertation, Linköping University (1997)

6. Fazlollahi, S., Mandel, P., Becker, G., Maréchal, F.: Methods for multi-objective investment and operating optimization of complex energy systems. Energy **45**(1), 12–22 (2012)
7. Horii, S., Ito, K., Pak, P.S., Suzuki, Y.: Optimal planning of gas turbine co-generation plants based on mixed-integer linear programming. Int. J. Energy Res. **11**(4), 507–518 (1987)
8. IBM ILOG CPLEX Optimization Studio V12.5.1: http://pic.dhe.ibm.com/infocenter/cosinfoc/v12r5/index.jsp
9. Ito, K., Yokoyama, R., Akagi, S., Matsumoto, Y.: Influence of fuel cost on the operation of a gas turbine-waste heat boiler cogeneration plant. Trans. ASME. J. Eng. Gas Turbines Power **112**(1), 122–128 (1990)
10. Iyer, R.R., Grossmann, I.E.: Optimal multiperiod operational planning for utility systems. Comput. Chem. Eng. **21**(8), 787–800 (1997)
11. Iyer, R.R., Grossmann, I.E.: Synthesis and operational planning of utility systems for multiperiod operation. Comput. Chem. Eng. **22**(7–8), 979–993 (1998)
12. Kuester, J.L., Mize, J.H.: Optimization Techniques with FORTRAN. McGraw-Hill, NewYork (1972)
13. Lozano, M.A., Ramos, J.C., Carvalho, M., Serra, L.M.: Structure optimization of energy supply systems in tertiary sector buildings. Energy Build. **41**(10), 1063–1075 (2009)
14. Lozano, M.A., Ramos, J.C., Serra, L.M.: Cost optimization of the design of CHCP (combined heat, cooling and power) systems under legal constraints. Energy **35**(2), 794–805 (2010)
15. Owen, J.H., Mehrotra, S.: On the value of binary expansions for general mixed-integer linear programs. Oper. Res. **50**(5), 810–819 (2002)
16. Papoulias, S.A., Grossmann, I.E.: A structural optimization approach in process synthesis–I: utility systems. Comput. Chem. Eng. **7**(6), 695–706 (1983)
17. Piacentino, A., Barbaro, C., Cardona, F., Gallea, R., Cardona, E.: A comprehensive tool for efficient design and operation of polygeneration-based energy grids serving a cluster of buildings, part I: description of the method. Appl. Energy **111**, 1204–1221 (2013)
18. Sheblé, G.B., Fahd, G.N.: Unit commitment literature synopsis. IEEE Trans. Power Syst. **9**(1), 128–135 (1994)
19. Van den Bosch, P.P.J., Honderd, G.: A solution of the unit commitment problem via decomposition and dynamic programming. IEEE Trans. Power Appar. Syst. **104**(7), 1684–1690 (1985)
20. Voll, P., Hennen, M., Klaffke, C., Lampe, M., Bardow, A.: Exploring the near-optimal solution space for the synthesis of distributed energy supply systems. Chem. Eng. Trans. **35**(1), 277–282 (2013)
21. Voll, P., Klaffke, C., Hennen, M., Bardow, A.: Automated superstructure-based synthesis and optimization of distributed energy supply systems. Energy **50**, 374–388 (2013)
22. Wakui, T., Yokoyama, R.: Optimal structural design of residential cogeneration systems in consideration of their operating restrictions. Energy **64**, 719–733 (2014)
23. Yokoyama, R.: Optimal operation of a gas turbine cogeneration plant in consideration of equipment minimum up and down times. Trans. ASME J. Eng. Gas Turbines Power **135**(7): Paper No. 071801, 1–8 (2013)
24. Yokoyama, R., Fujiwara, K., Ohkura, M., Wakui, T.: A revised method for robust optimal design of energy supply systems based on minimax regret criterion. Energy Convers. Manag. **84**, 196–208 (2014)
25. Yokoyama, R., Hasegawa, Y., Ito, K.: A MILP decomposition approach to large scale optimization in structural design of energy supply systems. Energy Convers. Manag. **43**(6), 771–790 (2002)
26. Yokoyama, R., Ito, K.: Optimal operation of a cogeneration plant in consideration of equipment startup/shutdown cost. Trans. ASME J. Energy Res. Technol. **121**(4), 254–261 (1999)
27. Yokoyama, R., Ito, K.: Optimal design of gas turbine cogeneration plants in consideration of discreteness of equipment capacities. Trans. ASME J. Eng. Gas Turbines Power **128**(2), 336–343 (2006)
28. Yokoyama, R., Ito, K.: Performance evaluation of gas turbine cogeneration plants using a design optimization tool: OPS-design. Proc. ASME Turbo Expo 2006: Paper No. GT2006-90611, 1–10 (2006)

29. Yokoyama, R., Ohkura, M., Wakui, T.: Robust optimal operation of a gas turbine cogeneration plant under uncertain energy demands. Trans. ASME J. Eng. Gas Turbines Power **137**(2): Paper No. 022001, 1–11 (2015)
30. Yokoyama, R., Shinano, Y., Taniguchi, S., Ohkura, M., Wakui, T.: Optimization of energy supply systems by MILP branch and bound method in consideration of hierarchical relationship between design and operation. Energy Convers. Manag. **92**, 92–104 (2015)
31. Zhou, Z., Liu, P., Li, Z., Ni, W.: An engineering approach to the optimal design of distributed energy systems in China. Appl. Therm. Eng. **53**(2), 387–396 (2013)

Demand Response Optimization Based on Building's Characteristics

Tomoshi Otsuki

Abstract Demand response (DR) is one of the technologies that targets for power control based on the cooperation between power suppliers and consumers. In the case of buildings' power control, it is important for buildings to achieve buildings' power reduction target more correctly under individual buildings' less burden. We suppose the framework of buildings' aggregator that collects the information for power reduction (NEGAWATT information) and requests each building to reduce demand to achieve buildings' power reduction target efficiently. In the framework, we first collect the NEGAWATT information based on buildings' characteristics, and then we make the demand response plans (DR plans) that meet the requirements of buildings. In this paper, we focus on the DR optimization techniques based on NEGAWATT information, and show two simulation results, one of which shows the aggregation effect in a simple simulation, the other of which shows the multiple scenario methods to deal with the DR optimization under uncertainty. These simulations show that DR utilizing NEGAWATT information is more efficient for demand-supply balance than conventional DR methods.

Keywords Demand-supply balance · Demand response · Optimization · Multiple scenario method

1 Introduction

We have researched the technologies for smart community that compatibly realizes efficient energy use for individual buildings with comfort and rational energy use for the community in total. In this research, we have developed clustered BEMS (Building Energy Management System) which bundles the buildings and has demonstrated the demand response (DR) experiments in Yokohama Smart City Project (YSCP [1]), one of the largest DR pilot projects in Japan. The clustered BEMS is one of the energy aggregators that aggregates the power demand of buildings, and aims to reduce the

T. Otsuki (✉)
Toshiba Corporation, 1, Komukai-Toshiba-cho, Saiwai-ku, Kawasaki 212-8582, Japan
e-mail: tomoshi1.otsuki@toshiba.co.jp

K. Fujisawa et al. (eds.), *Optimization in the Real World*,
Mathematics for Industry 13, DOI 10.1007/978-4-431-55420-2_10

total demand through a demand response plan (DR plan) for the individual buildings. The aggregator makes DR plans to meet the requirements of buildings and requests the buildings to reduce demand based on the plan in order to achieve the buildings' demand reduction target. In this paper, we focus on the DR optimization techniques that consider demand characteristics of individual buildings under uncertainty of demand, and also show the evaluation results by simulations.

The rest of the paper is organized as follows. In Sect. 2, we describe the framework to realize DR. Then, Sect. 3 demonstrates the simulation result which shows effectiveness of building aggregation. Moreover, Sect. 4 shows the more sophisticated simulation results to make the DR plan under uncertainty of demand. Finally, Sect. 5 describes conclusion.[1]

2 Demand Response Using NEGAWATT Information

Figure 1 shows the aggregation framework in which the aggregator collects sufficient DR amount from buildings. There are various types in belonging buildings, such as buildings that possess generators or batteries, or buildings that reduce demand by energy-saving efforts. Considering these building's characteristics, the aggregator aims to achieve the total demand reduction target. We suppose the total demand reduction target is set for each time slot (e.g. on an hourly basis) as shown in upper left part of the Fig. 1. This represents the situation that utility companies (or other operators that handle the power grid like transmission grid operators) try to solve the demand-supply balance problems by requesting demand reduction to buildings.

First, each building registers NEGAWATT information beforehand (Fig. 1a). Here, NEGAWATT indicates the proposal demand reduction amount by buildings. Then, the aggregator makes a DR plan to dispatch reduction target amount to buildings and requests the demand reduction to each building (=DR request) in order to achieve the buildings' demand reduction target (Fig. 1b). Note that the request amount to each building must be equal to or less than the NEGAWATT amount registered. Finally, each building does actual demand reduction according to DR request.

Through this framework, utility companies, the aggregator, and buildings all have benefits as follows:

- Utility companies can solve demand-supply balance problems with less cost by utilizing NEGAWATT than that by starting up generators.
- The aggregator can earn a lot of margins if it grasps each building's demand control ability correctly and controlling the total demand with high precision by less incentive cost.
- Each building can reduce demand with less burden and can get incentive in response to the actual demand reduction amount.

[1]Some parts of this article, some diagrams and some tables have already been published in [2, 3]. Copyright @ 2014 Information Processing Society of Japan and Copyright @ 2013 Toshiba Corporation.

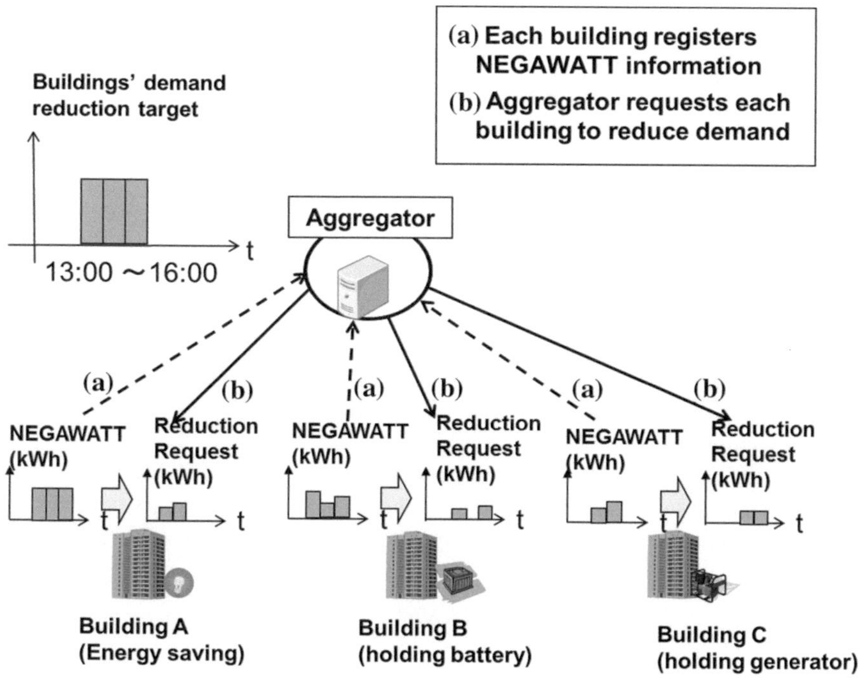

Fig. 1 An aggregation framework, in which each building registers NEGAWATT information first, then the aggregator requests each building to reduce demand

Thus, the goal of aggregator is collecting sufficient DR amounts effectively by considering building characteristics to reduce the total demand.

2.1 Buildings Characteristics

When the aggregator requests DR, the conventional methods, such as one in which the aggregator always requests each building to reduce prefixed same amount for all the slots, might be the burden of buildings, since buildings often need large power consumption in specified time (e.g. restaurants need much power in lunch or dinner time) or unexpectedly (e.g. by special events).

Thus, to minimize the burden of buildings, it is desirable to consider building's convenience depending on their characteristics, such as their sizes, types, or power-saving devices they possess as the following examples:

- Building *A* (practicing energy-saving): It can reduce demand by energy-saving at most 2 h. On the other hand, it can respond immediately to the DR request (e.g. 1 h before the beginning of actual demand reduction).

- Building B (holding buttery equipment): It can reduce 300 kWh per day in total based on the battery capacity. However, it requires making the charge-discharge schedule 5 h before the beginning of actual demand reduction.
- Building C (holding generation equipment): It can reduce demand by starting up generators. However, it requires making the generation plan 10 h before the beginning of actual demand reduction.

To meet these requirements, we model the features of NEGAWATT information for each building k as follows:

- Reduction capacity (kWh) A_{tk}: maximum demand reduction amount for each hour t.
- Unit cost (yen/kWh) C_{tk}: NEGAWATT cost for each hour t (the operating and fuel cost in buildings holding equipment, and the burden cost in buildings practicing energy-saving).
- Start-up time (hour) τ_k: necessary time to start up NEGAWATT resources.
- Maximum total hours (hour/day) M_k: total demand reduction hours per day.
- Maximum reduction amount (kWh/day) N_k: total demand reduction amount per day.

Table 1 shows the example of NEGAWATT information parameter for building A, B, and C in the above examples.

By considering these characteristics, we can reflect each building's requirement, which alleviates each building's burden more than the conventional method in which all the buildings are requested to reduce demand uniformly and for a long-time.

2.2 Related Work

DR can be categorized into two programs: time-based programs and incentive-based programs [4]. In time-based programs, customers are requested to reduce demand

Table 1 An example of NEGAWATT information

Buildings	A	B	C
Reduction capacity (kWh): A_{tk}	100	300	200
Unit cost (yen/kWh): C_{tk}	80	20	30
Start-up time: τ_k	1	5	10
Maximum total hours (hour/day): M_k	2	24	24
Maximum reduction amount: N_k	–	300	–

simply by changing power unit price. In incentive-based programs, on the other hand, customers are incentivized under certain agreed conditions depending on the customer's ability such as amount, response time, or stability of reduction. Both time-based and incentive-based programs have been realized in some US power markets or aggregators (See [5] for an example of time-based programs and [6] for incentive-based programs).

Similar to the DR planning problems, the unit commitment problem that determines the start-and-stop control for electric power systems in utility companies that balances the demand-supply gap of power by controlling multiple generators the utility companies hold is well-known. There are a number of solutions known as to the unit commitment problem such as Lagrange's method [7, 8], genetic algorithm [9], and stochastic programming [10]. However, since we generally do not use the demand side information on the unit commitment settings, these solutions cannot be directly extended to our problems.

3 DR Simulation I: Effectiveness of Aggregation

In this section, we show the aggregation effect by a simple example, evaluating "peak cut effect" by cooperation of buildings in using pseudo building demand data.

3.1 Problem Setting

Here, we evaluate the peak cut effect by making use of reduction capacity, by comparing the peak cut rate of the case only by individual efforts and the case by corporation based on optimization.

To simplify the situation, we hypothesize as follows:

- Individual buildings can reduce demand within reduction capacity, and reduce for three consecutive hours,
- Reduction capacity ($=A_{tk}$) is simply determined by 20 % of air conditioner demand plus 15 % of lighting demand (shown in Table 2), and
- Individual buildings reduce demand just as requested.

Then, we set up 7 building types in Japan and derive typical demand curves and energy composition from the public documents [11, 12] as shown in Fig. 2 and Table 2.

In Table 2, "building type" denotes the name of 7 building category types, and the "type rate" represents the demand percentage against total demand of all 7 buildings. Then, the columns of "air conditioner" and "lighting" denote the composition rate of conditioner and lighting demand against total demand of each building, respectively. Moreover, Fig. 2 shows the demand curve for each category building, whose peak demand is normalized at 100.

Table 2 Type rate, air conditioner rate, and lighting rate for each building type

Building type	Type rate (%)	Air conditioner rate (%)	Lighting rate (%)
Office	19	48	24
Whole sale/retail	25	48	26
Supermarket	14	25	24
Hospital	7	38	37
Hotel	23	26	31
Restaurant	4	46	29
School	9	7	69

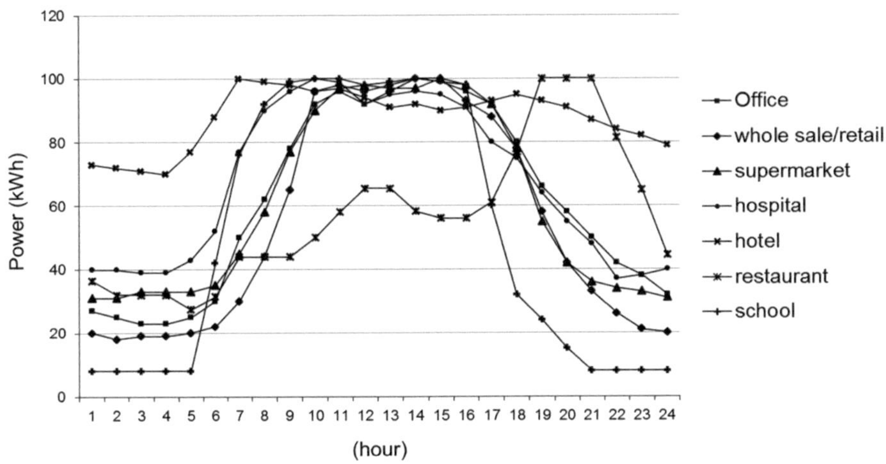

Fig. 2 An example of demand curves for each building type. Each curve is normalized such that the peak demand is at 100

3.2 Peak Cut Optimization by MILP

The mixed integer linear programming formulation of peak cut optimization is as follows:

$$\min \quad y, \tag{1}$$

$$\text{s.t.} \sum_{k \in K} \{E_{tk} - A_{tk}(x_{tk} + x_{t-1\,k} + x_{t-2\,k})\} \leq y, \ \forall t \in \mathscr{T}, \tag{2}$$

$$\sum_{t \in \mathscr{T}} x_{tk} = 1, \quad \forall k \in \mathscr{K}, \tag{3}$$

$$x_{tk} \in \{0, 1\}, \quad \forall t \in \mathscr{T}, \forall k \in \mathscr{K}. \tag{4}$$

- $\mathscr{T} = \{0, 1, \ldots, 23\}$: set of hours.
- $\mathscr{K} = \{1, \ldots, 7\}$: set of buildings.
- A_{tk}: reduction capacity of building k at t.
- E_{tk}: base demand of building k at t?
- y: variable: peak of the total demand.
- x_{tk}: 0–1 variable: building k starts actual demand reduction at t or not?

In this case, decision variables are the 0–1 variable x_{tk} representing building k starts actual demand reduction at t or not, and the continuous variable y representing the peak of the total demand. Each building's base demand E_{tk} is obtained by multiplying the value in Fig. 2 by type rates in Table 2. Reduction capacity A_{tk} is obtained by the following equation

$$A_{tk} = E_{tk} \cdot \{0.2 \times (\text{air conditioner rate}) + 0.15 \times (\text{lighting rate})\}, \tag{5}$$

where air conditioner rate and lighting rate are given in Table 2.

Equation 1 is an objective function that minimizes the peak of the total demand curve with Eq. 2 under the condition that each building can reduce for three consecutive hours with reduction capacity of each hour. Then, Eq. 3 represents that each building selects one of the slots to start actual demand reduction.

3.3 Optimization Result

Figure 3 shows the simulation results. The top table of the figure shows the result of the individual-saving case, minimizing each building's own peak demand, while the bottom table of the figure shows the result of the cooperation case, minimizing buildings' peak demand by cooperation of buildings. The cell with star indicates the peak time of each building, while the cells colored in gray show actual reduction timing for each building.

In the individual-saving case, reduction hours are set by three consecutive hours whose center hour corresponds to their peak time. On the other hand, in the cooperation case, reduction hours are calculated by Eqs. 1–4. Note that, in the cooperation case, each building does not necessarily reduce demand on its own peak, since they aim to reduce the buildings' peak demand.

Then, Fig. 4 shows the demand curve of each method, while Fig. 5 shows the peak values of these curves. As shown in Fig. 5, the peak reduction in the cooperation case is 5.1 % against the no-reduction case, while that in the individual-saving case is 1.5 %.

As a result, cooperation gives much greater peak cut effect in terms of the buildings' peak.

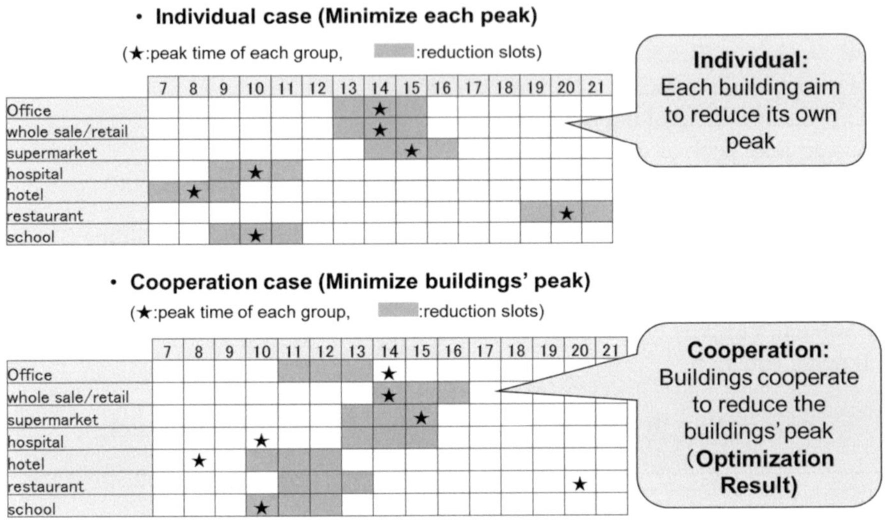

Fig. 3 Peak cut simulation results by the individual-saving case and the cooperation case

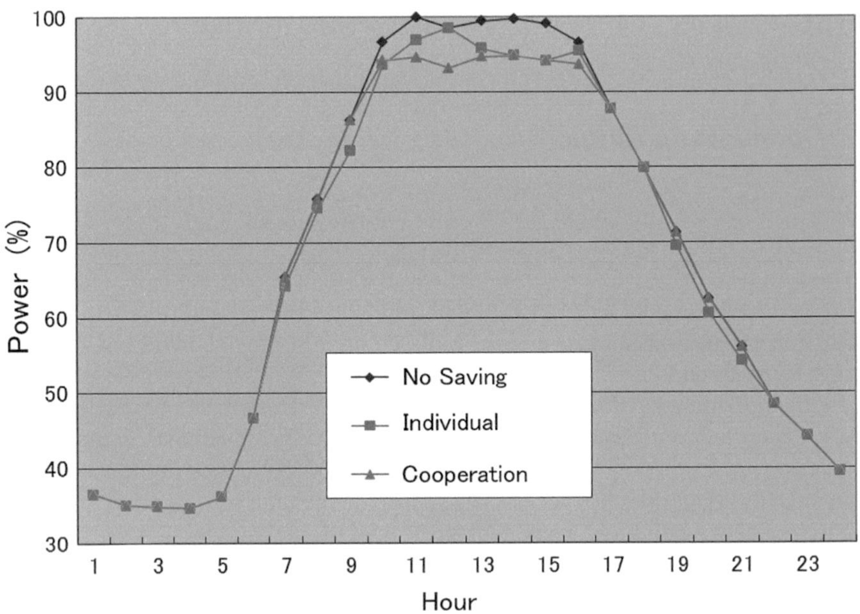

Fig. 4 Demand curves by the no-saving, individual-saving and the cooperation case. Demand power values in y-axis is normalized by the peak demand of the no-saving case

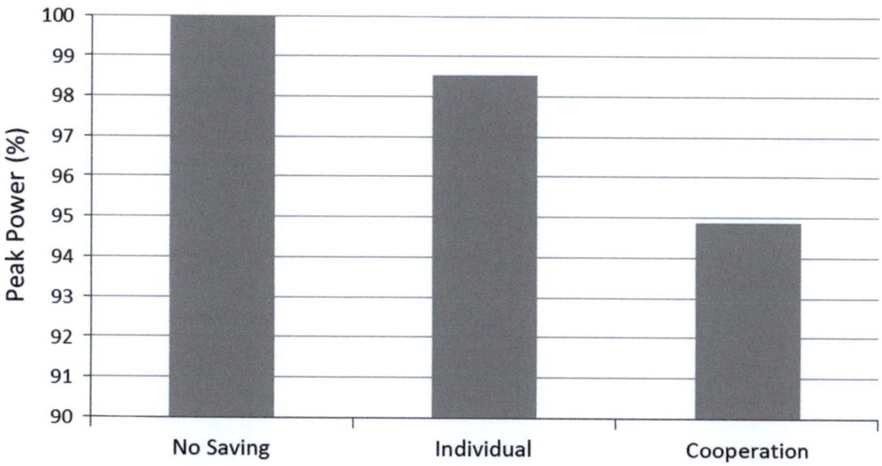

Fig. 5 Peak values in demand curves by the no-saving, individual-saving and the cooperation case. Peak values are normalized by the peak demand of the no-saving case

4 DR Simulation II: DR Planning Under Uncertainty of Demand Prediction

Generally, power demand fluctuates by emergent necessities or meteorological factors, so that it is difficult to predict demand accurately. Thus, the aggregator runs the risk of not achieving buildings' demand reduction target caused by demand prediction error as shown in Fig. 6.

In the example, we suppose buildings' initial demand prediction value, which corresponds to the buildings' base demand when no reduction would be conducted, is 200 kWh, and suppose committed buildings' demand reduction target amount is 40 kWh by reducing building's total demand to 160 kWh. Then, consider the case in which buildings' base demand is increased by 60 kWh, due to unexpected factors such as the increase in forecast temperature. In this case, to achieve the buildings' intended demand reduction target (that is, achieving 160 kWh), the aggregator has to reduce demand by 100 kWh in total, corresponding to the original buildings' demand reduction target (=40 kWh) plus unexpected increase amount (=60 kWh). Thus, unless the aggregator simulates such situations, the aggregator would fail to achieve the buildings' demand reduction target.

Thus, when we make DR plans, we need to deal with demand prediction errors caused by the fluctuation of the demand.

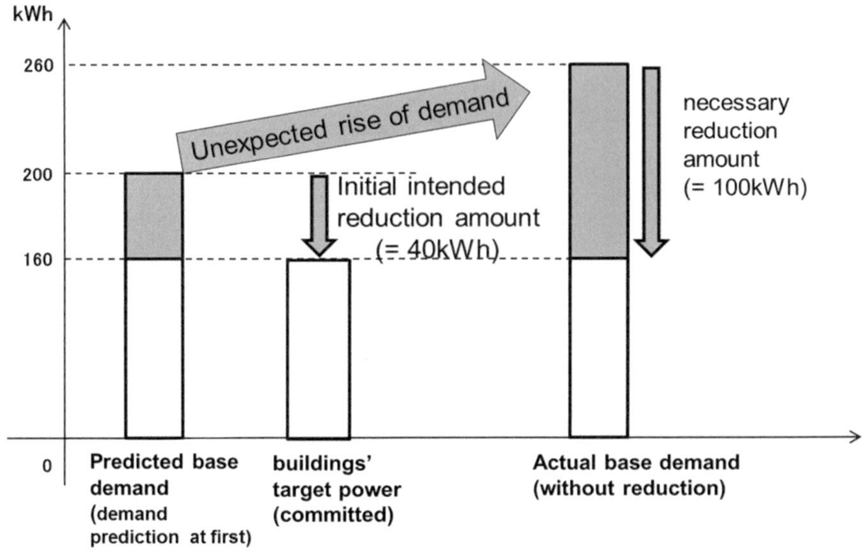

Fig. 6 An example in which demand prediction error causes the difficulty of achieving buildings' demand reduction target

4.1 Problem Setting

To deal with the demand fluctuation, we can make use of the start-up time of the NEGAWATT resources. Here, we take the strategy to make the hourly DR plan with the minimized cost under uncertainty. Therefore, the decision variables consist of the demand reduction amount and DR request timings to each building.

Moreover, we suppose that the cost of the aggregator is the sum of buying NEGAWATT cost and penalty fee as follows:

- NEGAWATT cost: cost for buying NEGAWATT from each building
- Penalty fee: penalty fee when the aggregator fails to achieve buildings' demand reduction target

Figure 7 shows the example of the hourly DR plan satisfying the buildings' NEGAWATT information in the Fig. 1. In this example, DR request timings, which mean the timings in which the aggregator inform each building of how large and when it should reduce demand, are set on 3:00–15:00, while actual DR execution timings, which mean the timings in which each buildings reduce demand actually, are set on 13:00–16:00 corresponding to 3 time slots.

In this DR plan, the aggregator requests the building A at 12:00 (and 13:00) to execute the 100 kWh DR on 13:00–14:00 (and 14:00–15:00), requests the building B at 10:00 to execute the 300 kWh DR on 15:00–16:00, and requests the building C at 3:00 (and 4:00) to execute the 200 kWh DR on 13:00–14:00 (and 14:00–15:00).

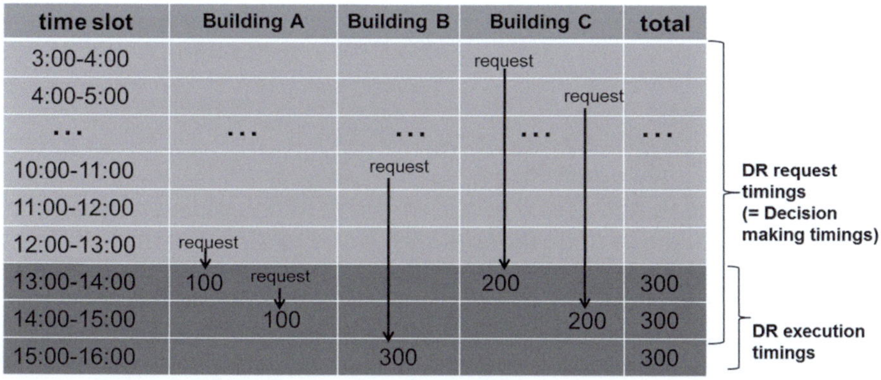

Fig. 7 An example of hourly DR plan in which the aggregator requests demand reduction to each building. DR request timings are set on 3:00–15:00, while actual demand reduction timings are set on 13:00–16:00 corresponding to 3 time slots

These requests achieve the 300 kWh buildings' demand reduction target and satisfy the constraints of each building's NEGAWATT information as follows:

- The total request hours of building $A(=2)$ is $\leq M_k(=2)$
- The total request amount of Building $B(=300)$ is $\leq A_{tk}(=300)$
- The difference of the DR request timing and the corresponding DR execution timing of A (:1 h), B (:5 h), and C (:6 h) are less than or equal to the start-up time τ_k, respectively.

Note that when a DR request timing to a building is set at 10:00, if the aggregator judges the request is unnecessary before 10:00, the aggregator can revoke the request.

4.2 DR Planning Optimization by Multiple Scenarios

In this section, we show how to solve this problem.

Consider the case that the demand prediction fluctuates as shown in Fig. 8. For simplicity, we consider 2-phase decision making at T_1 (e.g. at 8:00, 5 h before DR execution), and at T_2 (e.g. at 12:00, 1 hour before DR execution) to make the 13:00–14:00 DR plan. Note that, for building B, which has the NEGAWATT with its start-up time of 5 h, we can request reduction only at T_1, while for building A, which has the NEGAWATT with its start-up time of 1 h, we can request reduction both at T_1 and T_2. The aggregator can request DR for building A or B at T_1 based on the demand prediction at T_1, while it can request DR for both building A at T_2, based on the updated (and typically more precise) demand prediction at T_2.

Moreover, for simplicity, we suppose two possible prediction variations at T_1, one of which corresponds to the increasing demand prediction (variation 1), the other of which corresponds to the decreasing demand prediction (variation 2). In the

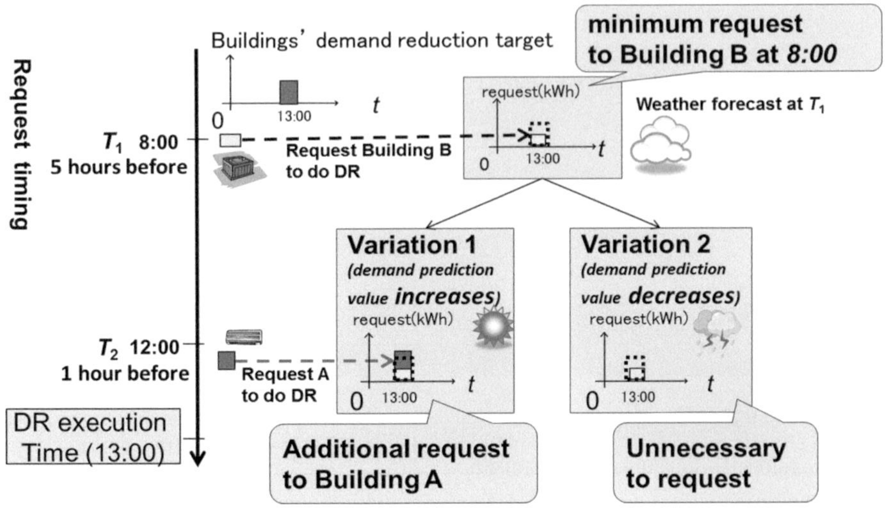

Fig. 8 An example of two variations in demand prediction

prediction variation 1, the aggregator needs more DR request amount than that at the plan at T_1 so it needs to make additional request for building A. In the prediction variation 2, on the other hand, since no additional request is necessary, the aggregator needs to request no additional reduction. Thus, if we can know the probability of these variations (for variation 1 and 2) beforehand, we can minimize the cost by considering these two scenarios. For example, making the DR plan with minimum DR request for building B at T_1 will seem to minimize the total cost.

As shown above, at individual decision making timing, the aggregator makes and updates DR plans for each building based on the demand prediction, which minimizes the cost under the NEGAWATT information.

Here, though we show the simple example with both the number of decision making timings and prediction variations are two, we have to consider more complex problems in realistic situations. As shown in the Fig. 9a, if we suppose the number of decision making timings and predictions equal to N and k, respectively, the number of total prediction variations are k^N, which leads to the combinatorial expansion that makes it very hard to solve.

To deal with making DR plans (e.g. at T_1) under uncertainty, we have some options.

Naively, we can replace reduction capacity value by the expectation values of demand prediction. However, simple use of expectation values cannot deal with the fluctuation correctly (See the later simulation results), especially in the case of the large variance in demand prediction distribution, since it drops the variance information.

As another method to solve this problem, we can possibly use the stochastic programming [13] that uses the stochastic variable directly to deal with the uncertainty

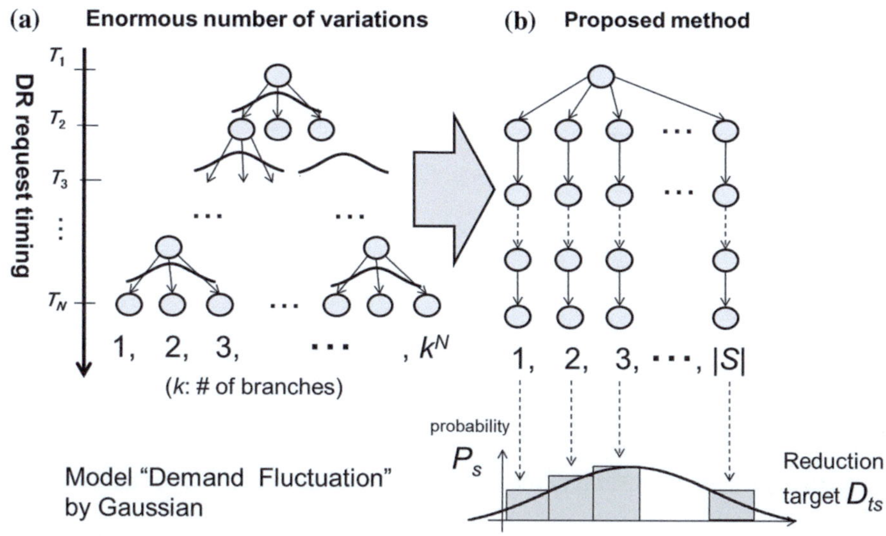

Fig. 9 An example that shows the combinatorial expansion of the prediction variations and the solution by multiple scenarios

of distributions. However, when we use the conventional methods such as L-shaped method [14], it seems to be hard to solve multiple-stage optimization in short time.

Therefore, in order to consider various possible situations effectively, we apply sample path optimization methods [15] (or scenario tree, scenario aggregation method [16]) in which we approximate the demand prediction distribution by multiple scenarios corresponding to the demand prediction variations in the future.

Here, we arrange representative $|S|$ scenarios based on the demand prediction represented by Gaussian, and approximate the demand prediction by these scenarios as shown in the Fig. 9b. As a result, we can get sub-optimum solutions for large-scale problems in short time.

4.2.1 DR Planning Optimization by MILP

Here, we explain the MILP formulation of the DR planning problem by using sample path optimization to make DR plan at t_0 (See [3] for details). Note that, this is the solution at current time t_0 and we need to solve this problem hourly and update the DR plan based on the updated demand prediction.

$$\min \sum_{s \in \mathscr{S}} P_s \cdot \sum_{t \in \mathscr{T}_c} \Big(\sum_{k \in \mathscr{I}_t} C_{tk} v_{tk} + \sum_{k \in \mathscr{J}_t} C_{tk} w_{tks} + R_t z_{ts} \Big) \tag{6}$$

$$\text{s.t.} \sum\nolimits_{k \in \mathscr{I}_t} v_{tk} + \sum\nolimits_{k \in \mathscr{J}_t} w_{tks} \geq D_{ts}(1 - z_{ts}), \forall t \in \mathscr{T}_c, \forall s \in \mathscr{S}, \tag{7}$$

$$0 \le v_{tk} \le A_{tk}x_{tk}, \qquad \forall t \in \mathscr{T}_c, \forall k \in \mathscr{I}_t, \tag{8}$$

$$0 \le w_{tks} \le A_{tk}y_{tks}, \qquad \forall t \in \mathscr{T}_c, \forall k \in \mathscr{J}_t, \forall s \in \mathscr{S}, \tag{9}$$

$$\sum_{t:t \le t_0+\tau_k} x_{tk} + \sum_{t:t \ge t_0+\tau_k+1} y_{tks} \le M_k, \quad \forall k \in \mathscr{I}_t, \forall s \in \mathscr{S}, \tag{10}$$

$$\sum_{t:t \le t_0+\tau_k} v_{tk} + \sum_{t:t \ge t_0+\tau_k+1} w_{tks} \le N_k, \quad \forall k \in \mathscr{I}_t, \forall s \in \mathscr{S}, \tag{11}$$

$$x_{tk} \in \{0, 1\}, \qquad \forall t \in \mathscr{T}_c, \forall k \in \mathscr{I}_t, \tag{12}$$

$$y_{tks} \in \{0, 1\}, \qquad \forall t \in \mathscr{T}_c, \forall k \in \mathscr{J}_t, \forall s \in \mathscr{S}, \tag{13}$$

$$z_{ts} \in \{0, 1\}, \qquad \forall t \in \mathscr{T}_c, \forall s \in \mathscr{S}. \tag{14}$$

Symbols are defined as follows:

- x_{tk}: 0–1 variable representing the aggregator requests reduction for building k at t or not.
- y_{tks}: 0–1 variable representing the aggregator requests reduction for building k at t or not in scenario s.
- z_{ts}: 0–1 variable representing the aggregator achieves buildings' demand reduction target at t or not in scenario s.
- v_{tk}: variable corresponding to the DR request amount for building k at t.
- w_{tks}: variable corresponding to the DR request amount for building k at t in scenario s.
- $\mathscr{T}_c$: set of hours at which the aggregator can request DR.
- $\mathscr{K}_t$: set of buildings that can reduce demand at t.
- $\mathscr{I}_t$: set of buildings in $\mathscr{K}_t$ that the aggregator has to request reduction before t_0.
- $\mathscr{J}_t$: set of buildings in $\mathscr{K}_t$ that the aggregator can request at t_0 or later.
- $\mathscr{S}$: set of current demand scenarios at t_0 corresponding demand prediction after t_0.
- A_{tk}: reduction capacity for building k at t.
- C_{tk}: unit cost for building k at t.
- τ_k: start-up time for building k.
- M_k: maximum reduction hours for building k.
- N_k: maximum reduction amount for building k.
- D_{ts}: buildings' demand reduction target at t in scenario s.
- P_s: probability of scenario s.
- R_t : penalty of the failure at t.

Here, decision variables are x_{tk},y_{tks},v_{tk},w_{tks} and z_{ts} as shown above. Note that the definition region of suffix t, k and s are $\mathscr{T}_c$, $\mathscr{K}_t(\equiv \mathscr{I}_t \cup \mathscr{J}_t)$ and $\mathscr{S}$, respectively. Moreover, x_{tk} and v_{tk} whose t satisfies $t < t_0$ is constant, since these values are already fixed by previously-conducted DR request.

Equation 6 is the objective function. The first term of the Eq. 6 represents NEGAWATT cost corresponding to the request at t_0 or earlier. On the other hand, the second and the third term of Eq. 6 represents the expectation of the NEGAWATT cost and penalty fee on multiple scenarios in the future, respectively. Thus, the Eq. 6 represents the minimization of expectation cost in totality.

Then, we describe the constraints. First, Eq. 7 represents that the total amount of DR request to each building is over or equal to the buildings' demand reduction target on each scenario. Then, Eq. 8 represents the capacity constraints for each building regarding to the previously-conducted DR request, while Eq. 9 represents those requested at t_0 or later. Finally, Eq. 10 represents that the total demand reduction time is less than or equal to the maximum total hours, while Eq. 11 represents that the total demand reduction amount is less than or equal to the maximum total amounts.

4.3 Simulation Settings

Then, we describe the simulation settings (See [3] for details). The summary of the simulation is as follows:

- Actual demand reduction time slots are 13:00–15:00, that is, $\mathscr{T}_c = \{13, 14, 15\}$ and buildings' demand reduction target each is 500 kWh.
- DR request time slots are 0:00–15:00, and at each slot we make DR plans by the following four methods and request reduction to each building if necessary.
- We suppose 8 buildings that execute energy-saving, hold batteries or generators.
- Penalty fee R_t is 20,000 yen for each slot.
- Model demand prediction fluctuation by Gaussian.
- Generate 100 datasets randomly.

Using this problem setting, we evaluate the expectation of total cost by comparing the following 4 methods:

- proposed method: the method using $|S|$ scenarios in Eqs. 6–14 (we set $|S| = 2 \sim 15$ and in each case we set $P_s = 1/|S|$ and set D_{ts} values by $1/|S|$ percent point of supposed Gaussian distribution),
- fix1: the method using just one scenario in Eqs. 6–14 (set $|S| = 1$ and set $D_{t1} = \mu + \sigma$),
- fix2: the method using just one scenario in Eqs. 6–14 (set $|S| = 1$ and set $D_{t1} = \mu + 2\sigma$), and
- perfect: Perfect information scenario replacing demand prediction scenario by actual demand value in the future (set $|S| = 1$ and set D_{t1} at actual demand value in Eqs. 6–14),

where μ and σ denote the mean and the standard deviation of the supposed demand prediction distribution.

Note that, fixed scenario methods are the simpler approximation of the demand prediction, by representing the demand prediction distribution only by one pessimistic scenario that means the aggregator needs to reduce power more than buildings' intended power reduction target. The fix1 and fix2 methods correspond to the situation in which the DR amount aggregator needs to reduce is more than buildings' intended power reduction target by σ and 2σ, respectively. Thus, the fix2 method supposes the more pessimistic scenario than the fix1 method. On the other hand, the perfect method corresponds to the method, in which we predict the demand perfectly, though it is impossible in realty. Thus, the perfect method gives the lower bound of the cost on this problem setting.

4.4 Simulation Results

Figure 10 represents the average cost and the average number of failures against the number of scenarios $|S| = 2 \sim 15$, in 59 datasets (out of randomly-generated 100 datasets) in which the perfect method achieves the buildings' demand reduction target for all three slots by replacing the predict demand values by actual ones. Here, average cost denotes the expectation value for the sum of the NEGAWATT cost plus penalty fee, while the number of failures denotes how many times the aggregator cannot achieve buildings' demand reduction target in three slots. Figure 10 shows that both of the average cost and the average number of failures tends to reduce as $|S|$ increases until near $|S| = 10$.

Then, Table 3 shows the average cost (and detailed cost by NEGAWATT cost and penalty fee) and the number of failures for the above four methods in 59 datasets. Note that the perfect method can achieve the penalty fee at 0 due to the definition of the perfect method. Moreover, the proposed method uses the best $|S| = 9$ case.

The result shows that proposed method achieves minimum cost and the minimum number of failures among 3 methods, excluding the perfect method. Then, comparing the fix1 and fix2 methods, the fix2 method, which has safer margin, achieves less penalty fee (corresponding to less number of failures) but needs more NEGAWATT cost than the fix1 method, so that the NEGAWATT cost and the number of failures show trade-off relationship. On the other hand, since the proposed method realizes the less number of failures in less cost, it achieves lowest total cost.

Table 3 Average cost and the number of failures by proposed methods compared with 3 methods

Solutions	Proposed methods ($|S| = 9$)	Fix1	Fix2	Perfect
Averaged cost(yen)	28399.4	32525.5	31464.2	23709.6
(NEGAWATT cost)	27810.1	25721.9	27392.8	23709.6
(Penalty fee)	589.3	6803.6	4071.4	0.0
Average number of failures	0.01	0.11	0.07	0.0

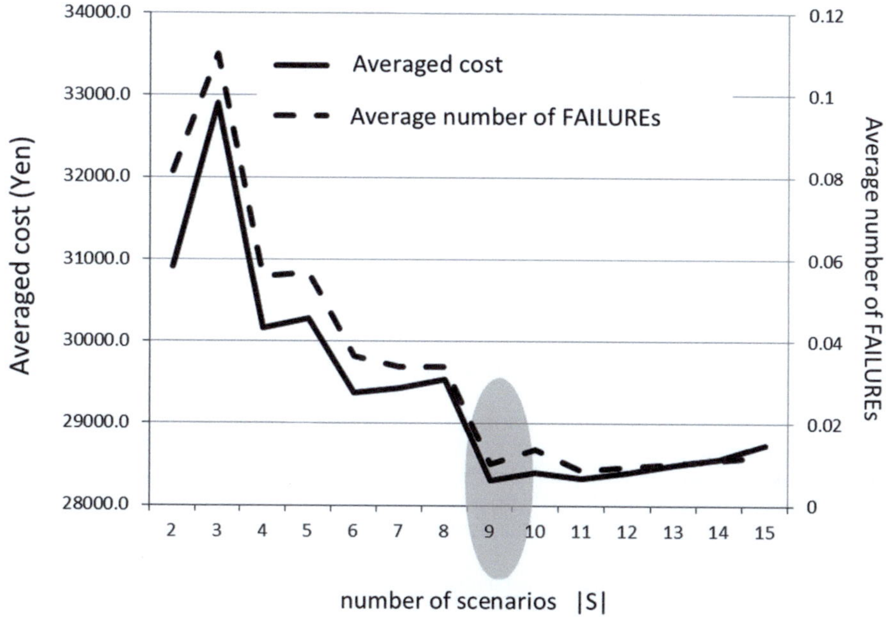

Fig. 10 Line plot of average cost and the average number of failures against the number of scenarios ($|S| = 2 \sim 15$) in the proposed methods

5 Conclusion

We introduce recent research activities for DR in two simulations. In the first simulation, we show that the cooperation of the buildings is efficient for peak cut effect of the total demand. Then, in the second simulation, we show the result in which we deal with the uncertainty of the demand prediction by multiple path methods. As a result, the proposed method achieves the lowest cost in our simulation settings. In the future, we improve the model and solution of DR planning to aim at the efficient utilizing of energy as buildings.

Acknowledgments This research is partially supported by the national project, the Next-generation Energy and Social System Demonstration Project, initiated by METI.

References

1. Japan Smart City Portal: http://jscp.nepc.or.jp/en/. Accessed on 9 Jan 2015
2. Otsuki, T.: Negawatt planning based on consumer's characteristics under uncertainty. Trans. Math. Model. Appl. Inf. Process. Soc. Jpn. (IPSJ-TOM), **7**(1), 44–52 (2014)
3. Otsuki, T., Aisu, H., Iino, Y.: DR optimization based on building's characteristics under uncertainty. Toshiba Rev. **68**(7), 23–26 (2013)

4. Goldman, C., Reid, M., Levy, R., Silverstein, A.: Coordination of energy efficiency and demand response, LBNL-3044E, Jan 2010
5. Carlson, B., Chen, Y., Hong, M.: MISO unlocks billions in savings through the application of operations research for energy and ancillary services markets. Interfaces **42**(1), 58–73 (2012). Jan
6. Schisler, K., et al.: The role of demand response in ancillary services markets. In: Transmission and Distribution Exposition Conference: 2008 IEEE PES Powering Toward the Future, PIMS 2008. No. 4517087 (2008)
7. Bard, J.F.: Short-term scheduling of thermal-electric generators using Lagrangian relaxations. Oper. Res. **36**, 756–766 (1988)
8. Muckstadt, J.A., Koenig, S.A.: An application of Lagrangian relaxation to scheduling in power-generation systems. Oper. Res. **25**, 387–403 (1977)
9. Tokoro, K., Masuda, Y., Nishino, H.: A planning method using genetic algorithm for large scale unit commitment problem. Central Research Institute of Electric Power Industry. Report No. R04018, Jan 2006
10. Shiina, T., Birge, J.R.: Multistage stochastic programming model for electric power capacity expansion problem. Jpn. J. Ind. Appl. Math. **20**, 379–397 (2003)
11. Energy saving standard format for small and medium-sized business: http://www.meti.go.jp/earthquake/electricity_supply/0513_electricity_supply_02_07.pdf. Accessed on 9 Jan 2015
12. Energy white paper 2010: http://www.enecho.meti.go.jp/about/whitepaper/2010pdf/. Accessed on 9 Jan 2015
13. Birge, J.R.: Stochastic programming computation and applications. INFORMS J. Comput. **9**, 111–133 (1997)
14. Van Slyke, R.M., Wets, R.: L-shaped linear programs with applications to optimal control and stochastic programming. SIAM J. Appl. Math. **17**(4), 638–663 (1969)
15. Plambeck, E.L., Fu, B.R., Robinson, S.M., Suri, R.: Sample path optimization of convex stochastic performance functions. Math. Program. **75**, 137–176 (1996)
16. Rockafellar, R.T., Wets, R.J.B.: Scenarios and policy aggregation in optimization under uncertainty. Math. Oper. Res. **16**(1), 119–147 (1991)